SCIENCE AND LITERATURE
A series edited by George Levine

Gaston Bachelard, Subversive Humanist

Texts and Readings

MARY McALLESTER JONES

The University of Wisconsin Press

The University of Wisconsin Press
114 North Murray Street
Madison, Wisconsin 53715

3 Henrietta Street
London WC2E 8LU, England

Copyright © 1991
The Board of Regents of the University of Wisconsin System
All rights reserved

5   4   3   2   1

Printed in the United States of America

Mary McAllester Jones's translations of "The Epistemological Break: Beyond Subject and Object in Modern Science" and "Towards a Non-Cartesian Epistemology" are printed with permission of Beacon Press, 25 Beacon Street, Boston, MA 02108. Her translations of "Mathematics and Poetry: On Lautréamont's Dynamic Imagination," "The Hand Dreams: On Material Imagination," "Isomorphic Images," "Narcissus," "Shelley's Prometheus: Images of Air," "Virginia Woolf's *Orlando*: The Image of the Tree," "Earth, Fire and Water: Images of the Smithy," and "Lamplight" are printed with permission of Dallas Institute of Humanities and Culture, 2719 Routh Street, Dallas, Texas 75201.

Library of Congress Cataloging-in-Publication Data
McAllester Jones, Mary
    Gaston Bachelard, subversive humanist : texts and readings
/ Mary McAllester Jones.
   220 pp.   cm.—(Science and literature)
    Includes translated extracts of Bachelard's works.
    Includes bibliographical references and index.
    1. Bachelard, Gaston, 1884–1962. 2. Literature and science.
3. Poetics. I. Bachelard, Gaston, 1884–1962. II. Title.
III. Series.
PN55.M34   1991
194—dc20
ISBN 0-299-12790-7   90-50649
ISBN 0-299-12794-X   CIP

For Robert

Contents

PREFACE ix

LIST OF ABBREVIATIONS xiii

Chapter 1 Bachelard in Perspective 3

Chapter 2 Polemics and Poetics: 1928–1932 15
   Extract I Mind versus Reality: Bachelard's Theory of Approximate Knowledge 21
   Extract II Duration, Instants, and the Absolute: Bergson or Einstein? 32
   Extract III On Consciousness and Time 36

Chapter 3 The New Scientific Mind: 1934 39
   Extract I The Epistemological Break: Beyond Subject and Object in Modern Science 46
   Extract II Toward a Non-Cartesian Epistemology 55

Chapter 4 Time, Consciousness, and Discontinuity: 1936 61
   Extract I (Cogito)^3 65
   Extract II On Poetry and the Dialectics of Duration 71

Chapter 5 Psychoanalyzing Reason: 1938 77
   Extract I Epistemological Obstacles 81
   Extract II The Teacher and the Taught 85

Chapter 6 Dynamic and Material Imagination: 1938–1948 91
   Extract I Mathematics and Poetry: On Lautréamont's Dynamic Imagination 98

Extract II	The Hand Dreams: On Material Imagination	102
Extract III	Isomorphic Images	107

Chapter 7 Reading Poetry: 1938–1948 — 112

Extract I	Narcissus	117
Extract II	Shelley's Prometheus: Images of Air	122
Extract III	Virginia Woolf's *Orlando*: The Image of the Tree	126
Extract IV	Earth, Fire, and Water: Images of the Smithy	129

Chapter 8 Applied Rationalism: 1949–1953 — 135

Extract I	Cogitamus: On the Psychology of Coexistence	142
Extract II	The Divided Subject	147

Chapter 9 Unfixing the Subject: 1957–1961 — 155

Extract	Lamplight	157

Chapter 10 Gaston Bachelard: Subversive Humanist — 164

APPENDIX: ENGLISH TRANSLATIONS OF BACHELARD'S WORK — 181

NOTES — 183

REFERENCE LIST — 189

INDEX — 197

Preface

Gaston Bachelard's work now attracts increasing attention in Britain and the United States, yet it is too often known only at second hand and by hearsay because few of his books on science, in particular, have been translated into English. The two colloquia on Bachelard held in London to mark the centenary of his birth, one in January 1980 at the City University (organized by the British Society for Phenomenology), the other in May 1984 at the French Institute, made it very clear that this lack of translation does not reflect lack of interest. My book was undertaken in this context, its principal purpose being to give the reader first-hand knowledge of what Bachelard himself wrote, setting his work on science alongside that on poetry. I have not however limited myself to offering a selection of Bachelard's texts. What I have called my "readings" of these texts are an integral part of this book, developed in response to two particular problems Bachelard's work poses, namely its frequent difficulty, stemming especially from his language and style, and its range, the interconnection of science and poetry over the years. These two problems may well explain why relatively little of the work of such a major thinker has been translated (details of published translations are given in the Bibliography), and why—with one exception—each of the translations available is by a different person. This last point, while understandable, is to be regretted, leading as it does to some inconsistency and to the loss of Bachelard's pattern of meaning, where words and ideas echo from book to book. I have tried to be consistent in my translating and alert to these echoes, and with this aim in view, all the translations from Bachelard here are my own. Similarly, for consistency's sake, I refer to all Bachelard's works by their French titles, giving a translation in the few cases where they may not be immediately understood.

Bachelard was a prolific writer, so that deciding which texts to include and which omit is inevitably a personal and partial matter. While limiting the texts translated to what can be agreed are his major books, I have drawn on a large number of his articles, conference papers, and books in my discussion. My selections have been guided by three considerations: first, to show the main lines of his thought and its development; second, to indicate the intellectual context in which this development takes place; third, to suggest why Bachelard remains relevant today. This last point is to me and, I believe, to others the most important one; it therefore provides the title and the argument of my book. Many of those English-speaking readers who are now drawn to Bachelard first encountered him through Michel Foucault and other participants in what has come to be known in France as the "humanist controversy" of the mid-1960s, for whom Bachelard was the inventor of the "epistemological break," the proponent of discontinuity. Bachelard, like many recent thinkers in France, rejected the primacy of lived experience and the conception of the founding, sovereign subject. He did not, though, believe that "man is dead." He redefined man; he reinterpreted the relationship between subject and object, examining human creativity in both science and poetry and placing considerable emphasis on language. Consequently, Bachelard's ideas are of far more than simply historical interest: his subversive humanism, as I have called it, matters now, and perhaps more than ever, to all of us.

Bachelard was always a polemical thinker, believing, as he declared in *La Philosophie du non* (1940), that "two people must first contradict each other if they really wish to understand each other. Truth is the child of argument, not of fond affinity" (134). It is in this spirit that I have discussed other books on Bachelard, and I hope through this polemic to have shown something of the "truth" of Bachelard. If I am critical of others, it is because Bachelard's texts suggest flaws in their argument. My own readings must, in their turn, be measured by these texts.

In translating, my aim has been fidelity both to Bachelard's French and to the English language. There is at times a degree of conflict between these, for Bachelard's use of French can be idiosyncratic. To take just one fairly simple example, the English-speaking reader may well be surprised by Bachelard's frequent use of "rationalism" and "rationalist," which he chooses in preference to "rationality" and "rational." *Rationalisme* and *rationalité, rationaliste* and *rationnel* have the same distinct meanings in French as in English. This distinction needs to be maintained in the translation, however awkwardly "rationalist," for instance, may strike the English ear. Where there is a conflict of loyalties, the translator's first duty, it seems to me, is to Bachelard.

Preface

I am indebted to Bachelard's publishers for kind permission to translate these extracts from his works: Presses Universitaires de France (*La Dialectique de la durée, Le Rationalisme appliqué*); Editions Stock (*L'Intuition de l'instant*); Messrs J. Vrin (*Essai sur la connaissance approchée, La Formation de l'esprit scientifique*). My own English translations of two extracts from *Le Nouvel Esprit scientifique* ("The Epistemological Break: Beyond Subject and Object in Modern Science", and "Towards a Non-Cartesian Epistemology") are printed with the permission of Beacon Press, 25 Beacon Street, Boston, MA 02108. My translations of extracts from *Lautréamont, L'Eau et les rêves, L'Air et les songes, La Terre et les rêveries de la volonté, La Terre et les rêveries du repos,* and *La Flamme d'une chandelle* are printed with the permission of the Dallas Institute of Humanities and Culture, 2719 Routh Street, Dallas, Texas 75201. I am most grateful to both publishers for their permission; details of the translations they have published are given in my Appendix.

A first version of Chapter 1, "Bachelard in Perspective," was published as "Bachelard Twenty Years On: An Assessment" in *Revue de Littérature Comparée* 58, no. 2 (April–June 1984): 165–176. My translation of Bachelard's discussion of Shelley in Chapter 7 has been published in a chapter, "On Science, Poetry, and the 'honey of being': Bachelard's Shelley," in *Philosophers' Poets*, ed. David Wood, Warwick Studies in Philosophy and Literature (London: Routledge, 1990). I am indebted to both editors for kind permission to use this material here.

I gratefully acknowledge the help and support of many friends and colleagues who have discussed this book with me over the years. Dr. Margaret Lowe first encouraged me to embark on it, but sadly did not live to see it completed. Particular thanks are due to Professor Colin Smith, formerly Professor of French Studies in the University of Reading, for kindly taking the time to read and comment on the manuscript. Last, to my husband Robert Jones for his understanding, his patience, and his proofreading, I wish to dedicate this book.

List of Abbreviations

Full details of publication are provided in the Reference List.

ARPC	*L'Activité rationaliste de la physique contemporaine*
AS	*L'Air et les songes*
DD	*La Dialectique de la durée*
DR	*Le Droit de rêver*
ECA	*Essai sur la connaissance approchée*
EEPC	*L'Expérience de l'espace dans la physique contemporaine*
Eng. rat.	*L'Engagement rationaliste*
ER	*L'Eau et les rêves*
Ét.	*Études*
FC	*La Flamme d'une chandelle*
FES	*La Formation de l'esprit scientifique*
Geneva	"La Vocation scientifique et l'âme humaine"
Ghent	"La Psychanalyse de la connaissance objective"
II	*L'Intuition de l'instant*
L	*Lautréamont*
MR	*Le Matérialisme rationnel*
NES	*Le Nouvel Esprit scientifique*
PCCM	*Le Pluralisme cohérent de la chimie moderne*
PE	*La Poétique de l'espace*
PF	*La Psychanalyse du feu*
PN	*La Philosophie du non*
PR	*La Poétique de la rêverie*
RA	*Le Rationalisme appliqué*
TRR	*La Terre et les rêveries du repos*
TRV	*La Terre et les rêveries de la volonté*
VIR	*La Valeur inductive de la relativité*

Gaston Bachelard, Subversive Humanist

Chapter 1

Bachelard in Perspective

> Discoveries made about the structure of space and time always react on the structure of the mind. Other kinds of discovery enrich human knowledge without affecting its basis. However, anything to do with conceptions of space will suggest very different ways of constructing knowledge. The discovery of America brought only a few names of rivers and hills, and a small number of geographical rarities like the Niagara Falls; it was not, in the end, an event for the human mind. But with it came the inference that the Earth is round, and heaven and thought, heart and reason, were thrown into disarray.
> —*L'Expérience de l'espace dans la physique contemporaine*

Bachelard died in October 1962, leaving a rich and singular legacy: some ninety publications in all, with twenty-three books—twelve on the philosophy of modern science, two on time and consciousness, nine on poetic imagination—published between 1928 and 1961, and a tenth book on poetry left unfinished when he died. Since then, many have expressed their debt to Bachelard's books on poetic images. Indeed, Bachelard is generally held to have inspired the "new criticism" of 1965 and after. His epistemology has been equally seminal. Georges Canguilhem, for instance, in *Idéologie et rationalité*, referred to the "lessons learned from Gaston Bachelard" which had "inspired and fortified his young colleagues" (1977:9). Yet soon after this, Vincent Descombes in his book *Modern French Philosophy* mentioned Bachelard only three times (1980:89, 91, 120). Descombes limited his definition of "contemporary French philosophy" to "that which was spoken about" (1980:2). If Bachelard was left out, it was because apparently in 1978 he was no longer talked about. Canguilhem, though, was talking about him in 1977, and Bachelard was and still is widely read in France, an established part of the philosophy syllabus at school and university: the path to the *agrégation de philosophie* leads through Bachelard. France's young philosophers have not in fact lapsed into uncharacteristic silence. On the contrary, in the mid-seventies Bachelard became a controversial figure for the younger generation in particular. The Marxists Dominique Lecourt and Michel Vadée disagreed over him in books published in 1974 and 1975. In 1977, Jean-Pierre Roy, wearing the colors of Althusser, Derrida, and semiotics, berated him for his "unscientific" approach to literature, for betraying "modern scientificity" (37). These three passionate, noisy books make Descombes's silence all the more puzzling.

Some skeleton lies perhaps in the Bachelardian cupboard that is best kept hidden. We must be bold, open the door, and have a good look for ourselves. We find not a skeleton but man, vigorous and very much alive, man who through the power of his reason and his imagination creates "a new nature." In science and in poetry, Bachelard believes, "the world is conditioned by man's provocation" (*ARPC* 141). This is the sort of statement that now gets Bachelard into trouble. He was a humanist, and in France humanism has become inadmissible. Structural linguistics in particular has challenged the long-held view of man the creator, constituting and controlling his language, and through language, his experience. The last twenty years in France have seen an intellectual revolution, the dethronement of the sovereign subject. Consequently, Bachelard's humanism may now seem wrongheaded and outdated. If humanism necessarily implies idealism, if it argues that man is all-powerful, autonomous, the origin of all experience, then of course Bachelard's humanism would be an embarrassing legacy. Humanism need not though rest on idealism. I wish to show that Bachelard's humanism does not, that it rests on a conception of man decentered, transcended by something beyond his control, yet paradoxically neither denied nor destroyed by this transcendent "other," but rather nourished and sustained by it. Bachelard's is a subversive humanism. In his work both on science and on poetry, he undermines traditional views of the subject every bit as effectively as recent thinkers in France. His adversaries and theirs are the same, yet the conclusion he draws is very different. Subversively, he reinvents man, against idealism, beyond conventional notions of subject and object.

The matrix of Bachelard's thought is twentieth-century science, the "new scientific mind" which he dates from 1905, from Einstein's special theory of relativity. His twelve books on modern science examine its impact on philosophy, showing how science has undermined our familiar epistemologies, so that neither rationalism nor realism, idealism nor materialism will serve as philosophies adequate to twentieth-century science. The year 1905 saw the break not just with all previous science, but with all previous philosophy. Bachelard's notion of the "epistemological break" is probably what is best known and most widely quoted from his work, yet those who borrow it—Foucault, Derrida, Althusser—fail to see that this epistemological break brings humanism in its train, a humanism which, in its turn, breaks with traditional humanism. This is, I realize, a controversial assertion on my part, yet it seems to me a conclusion we must draw and it is therefore fundamental to my argument. My grounds for it are as follows. For Bachelard, what is new about twentieth-century science is that "scientific progress always reveals a break, or better perpetual breaks, between ordinary knowledge and scientific knowledge" (*MR* 207). It is not simply that scientific knowledge no longer has first and foremost an empirical basis. Far more important, the rational basis of modern science has broken with reason as we practice it not just in "ordinary knowledge" but in deduction. Modern science is characterized by the primacy of reason, but this is not Cartesian or Kantian reason; it has broken with "good sense" and, as we shall see, with deductive logic. This brings us to Bachelard's humanism. Both Descartes and Kant understood the rational subject in terms of this notion of reason as a priori and deductive. If, as Bachelard argues, reason in science is not like this, then consequently the rational subject is not as Descartes and Kant conceived him; he is not the unchanging center of all knowledge and experience. Science is the matrix of Bachelard's thought, and from it springs a new conception of man, of consciousness, and, in due course, of poetry.

What is this new reason at work in modern science? Why does Bachelard argue that in consequence of this "revolution in reason" the twentieth century has seen what he calls a "psychic revolution," a "mental revolution" (*Eng. rat.*, 7–12)? Why in promoting this revolution does he subvert not only Descartes and Kant, but Bergson, Husserl, and Freud? And why explain modern mathematics by a quotation from Mallarmé? Why regard reading Valéry as an equally effective way of experiencing the "psychic revolution" that marks our age? These five questions, even before we answer them, show something of Bachelard's humanism.

"In modern physics and its creations, we confront a new nature, a nature produced by man's instruments, and thought by man in the perspective of human history" (*ARPC* 141). This is surely a dramatic humanism, when man creates not just science but nature itself. "Indeed, we can with complete confidence now speak of the creation of phenomena by man," Bachelard writes in *La Formation de l'esprit scientifique* (249). As an example of this, he gives the electron: "the electron existed before twentieth-century man. Yet before twentieth-century man, the electron did not sing. Now in the triode valve the electron sings. This phenomenological realization occurred at a precise point in mathematical and technical maturity" (*FES* 249). While Bachelard underlines the conjunction of mathematics and technology here, he does tend to give precedence to the former. For instance, in his *Essai sur la connaissance approchée*, he describes what he calls "the progressive mathematization of technology" (157), and in *Le Nouvel Esprit scientifique* stresses that in modern science "instruments are just materialized theories," the phenomena that "come out of them," as he puts it, bearing "on all sides the mark of theory" (see below, 54). Of course, mathematics and technology together produce phenomena. Nevertheless, Bachelard draws attention to those occasions when "suddenly, these efforts of mathematization are so successful that reality crystallizes along the axes provided by human thought, and new phenomena are produced" (*FES* 249). Thus, "mathematics forms the axis of discovery, and only mathematical expressions allow us to think phenomena" (*NES* 58). This mathematics, Bachelard argues, is nondeductive, non-Euclidean. It was non-Euclidean geometry and Riemann's tensor calculus in particular that made relativity theory possible, so inaugurating the "new scientific mind." Twentieth-century science is marked by what Bachelard calls the "non-Euclidean revolution" (*NES* 28), and the "Riemannian revolution" (*ECA* 28).

Two characteristics signal this "revolution in reason," and consequently shake the foundations of the rational subject.

First, it is the end of reason as a closed, a priori, deductive system. Einstein's theories were not deduced from Newton's: "relativistic astronomy does not in any way derive from that of Newton" (*NES* 45). It is the end therefore of reason as a closed system of necessity. Second, reason is no longer governed by what Bachelard calls "the ideal of identification" (*Eng. rat.*, 12). "The real meaning of the Riemannian revolution" lies, he says, in the fact that Riemann "understood that mathematical functions are better defined by their differential equations than by a whole" (*ECA* 28). More generally, Bachelard notes the increasing importance of the differential calculus in mathematics, "its power to diversify" (*Eng. rat.*, 110–11). Non-Euclidean geometry and indeed all modern science are characterized by what he calls "the rationalism of differentiation" (*Eng. rat.*, 130–31). All this is indeed revolutionary, but why should it matter to us? It is important because it forces us to revise our conception of the rational subject, of the relationship between reason and reality, subject and object.

Let us consider the rational subject first. The new mathematics was constructed by Lobachevsky, Bolyai, Riemann: it is men who make difference. Even more subversively, Bachelard argues that men are made different by difference, that the mathematician is changed by his mathematics: "psychologically speaking, you cannot fail to note the reaction of the mathematical tool on the user of that tool" (*NES* 59). Moreover, it is not only the mathematician who is psychologically changed, but all of us, for if we read about modern science, we shall derive what Bachelard calls "psychological benefit" (*NES* 88), our minds growing more agile, more alert, dynamic, creative (*NES* 55–58, 65, 182–83), and this is what he terms the "psychic revolution" that accompanies the "revolution in reason." This desire not just to instruct us but to change us is an important aspect of Bachelard's books on science. If, for example, reading Bachelard, we try to grasp the dialectics of matter and energy in microphysics, the dualism of waves and particles, we shall learn to maintain difference, to handle complexity, and be shaken out of the reductive, identity-ridden habits of ordinary life and thought. Plainly, Bachelard does not see the rational subject in Cartesian terms. In the last chapter of *Le Nouvel Esprit scientifique*, he sets out a "non-Cartesian epistemology," an epistemology that is not *against* but rather *beyond* Descartes. He keeps the notion of the *cogito,* but refuses its permanence (56, below). We can no longer say "I think therefore I am" but rather "I think difference, therefore I become different, and being different, I think new differences." Bachelard subverts Cartesian rationalism, and consequently he subverts Kantian idealism. If our minds are changed by scientific knowledge of the world, then we can no longer argue with Kant that the laws of the world conform to the laws of man's mind. Bachelard therefore proposes what he calls "non-Kantianism," an "open Kantianism," idealism *beyond* Kant, a "discursive idealism" which he defines as "the clear reconstruction of the self in confrontation with the not-self . . . a sequence of essentially different constructions" (*Ét.*, 92). The rational subject is no longer sovereign, no longer autonomous, identical, and unchanging, but rather transcended, upheld, created and recreated by something other than itself, by the "not-self," by the discursive, dialectical, dynamic interrelationship between reason and reality.

All this raises the question of the origin of difference, of whether it is a property of the mind, the product of intricate mathematical pattern making, or a property of matter, of reality. Mathematics is the instrument of modern science, and more than this, mathematics is a language. This is a constant theme in Bachelard's books on science. Given that it is a conception of language that has led recent thinkers to decenter the subject, it is surely of considerable interest to find Bachelard discussing the language of mathematics in terms of difference, just ten pages into his first work, *Essai sur la connaissance approchée*, his doctoral thesis of 1927. An equation, he suggests here, is a structure of difference, the equals sign in fact establishing difference between the known and unknown. He goes on to quote an article published in 1914 by J. H. Rosny: "every word expresses first of all a differentiation, otherwise it would be confused with all other words," adding his own interpretation, namely that "the more complex thinking is, the more clearly differentiated are its terms" (22; *ECA* 21). The language of mathematics, like all language, is a structure of difference; its symbols and operators are arbitrary and autonomous. The mathematician is therefore limited by this language, whose rules are not of his choosing. However, Bachelard does not argue that the subject is decentered by language, that difference belongs only to language. The context of his thought is "these burgeoning new languages" (*NES* 11; see below, 50), the creation by men of new mathematical languages, the "revolution in reason," or, as he also puts it, "the poetic strivings of mathematicians" (*NES* 35). The mathematician like the poet is in language, as we all are. But the mathematician and the poet, unlike most of us, open and renew language. As indeed Bachelard so clearly demonstrates, "the language of science is in a state of permanent semantic revolution" (*MR* 215). This semantic revolution is clear proof that language and user interact.

Bachelard is convinced that this interaction occurs because the user is saying something about reality. Mathematics creates and *realizes* the phenomena of modern science, but he argues that mathematics is a "stratagem" designed to capture reality: "from now on, reason will be working with variables which are obviously phenomenal" (*Eng. rat.*, 110–11). Mathematics creates difference in response to a rich reality (*Eng. rat.*, 116, 118–19). Mathematics is *nourished* by "experimental matter" (*Eng. rat.*, 115; *NES* 59). Writing of crystals, for example, Bachelard declares that "in the physical phenomenon, we find reasons for enriching and continuing mathematical thought" (*Eng. rat.*, 118). Mathematics and experiment, reason and reality, subject and object are therefore reciprocal and interdependent. Mathematics speaks the language of an endlessly rich, elusive reality, or better, of possibility: "does not the memory of the beautiful symbols of mathematics, where reality and possibility are conjoined, call to mind Mallarmé's images?" (*NES* 60). He quotes from Mallarmé's *Divagations*: "we ought never to overlook in our minds any of the possibilities that flutter round the figures of poetry, for however unlikely this may seem, they belong to the original, to the source of the figure," adding that "in the same way, pure mathematical possibilities belong to the real phenomenon, however unlikely this may be, given the first lessons we learn from immediate experience." This unexpected reference to Mallarmé helps us to understand that through language, or more precisely through new language that has broken with ordinary language, the mathematician, like the poet, explores possibility.

Mathematician and poet each sustain possibility, and moreover, each is sustained by it. Bachelard develops as a consequence of his epistemology a conception of consciousness which is non-Bergsonian and non-Husserlian. Husserl's conception of intentionality is too static, and Bachelard seeks to adapt it, arguing in his second series of books on science (1949–53) that, given the interdependence of reason and reality in modern science, the "rationalist consciousness" is dynamic: I am conscious of something other than myself, and this "other" changes me. This surely confirms both Bachelard's humanism and his refusal of idealism. The same is true of his long polemic with Bergson. Bachelard wrote two books on time and consciousness against Bergson, *L'Intuition de l'instant* (1932) and *La Dialectique de la durée* (1936), ending the latter with what one might call his consciousness of a poem, with a description of how he reads Valéry: "if we speak soundlessly, and allow image to follow image in quick succession, so that we are living at the meeting point, the point of superimposition, of all the different interpretations . . . reality will, in this way, be enfolded and adorned by the rich garment of conditionals. In place of the association of ideas there comes the ever possible dissociation of interpretations" (150). Poetry, like mathematics, is a structure of difference. Reading a poem is consciousness of difference, differences we create as we read, difference created in us by our reading. Consciousness for Bachelard is consciousness of the imbrication of subject and object. He therefore refuses Bergson's distinction between the "superficial self" and the "deep self," he refuses the idea of duration as continuity. Instead, consciousness is of difference.

Without the world, there is no consciousness, no human being, no human becoming. Bachelard adapts Bergson's "life force" to make of it what we may call an "intellectual force." Science requires us to think in new ways; it has turned us into creatures capable of profound intellectual change. "Because of the mental revolutions that necessarily accompany scientific inventions, mankind is turning into a mutating species, or to put it more precisely, into a species that needs to mutate, that suffers if it does not change. From an intellectual point of view, man needs to need" (*FES* 16). For Bachelard, difference is an ontological necessity. Since ordinary life is under the rule of identity, the only way we can experience and sustain this difference in ourselves is by thinking about science or, alternatively, by reading a poem. As he will write in *L'Air et les songes*, "reason in silence, read in silence . . . these are the primary factors in human becoming" (278).

Well before the publication of his books on poetic images, Bachelard places a high value on poetry, this value being closely bound to his conception of the relationship between man and the world that grows out of his work on science. As we turn from science to poetry, we must remember that these books were written when Bachelard held the chair in the history and philosophy of science at the Sorbonne, that his work on science and on poetry is interwoven. Science and poetry are, of course, distinct and separate activities, yet for Bachelard there is a common factor. Equation and image both break with everyday experience, both break with *homo faber*, both make us, as he puts it, *homo aleator*, the explorers of possibility in the world and in ourselves (*NES* 119). Bachelard's work does not present a diptych of "scientificity" and humanism, materialism and "ideology," science and "fable," as is now commonly said. Curiously, many who admire his work on science have imposed on it their own materialist preoccupations, ignoring its real lesson, which is, quite explicitly, that we must rid ourselves of all preconceptions, rationalist or realist, which blind us to the facts, to science as it is now, beyond the simple divisions of reason and reality. Yet while Bachelard's books on science have a pedagogical aim— he indeed describes himself as "more of a teacher than a philosopher" (*RA* 12)—he refuses the role of teacher in his books on poetry and has hard things to say about teachers of literature. "What a bad literature teacher I would have made," he declares in *La Flamme d'une chandelle* (105), congratulating himself! And along with this, he rejects the role of literary critic, for literary criticism turns literature into "a never-ending literature lesson" (*DR* 177). This comes as something of a shock, given Bachelard's reputation in modern French criticism. Paradoxically, his dislike of teachers and critics of literature stems from the same wariness with regard to theory that made him a teacher of science, from the same desire to attend to the facts. Our shock at Bachelard's anticritical stance is, in a way, a measure of his success, for it is largely thanks to him that criticism is now understood differently, not as dogmatic but as attending to the facts, to the text itself. Hence his lasting importance whenever literary theory threatens to overwhelm our own experience of reading a text.

Bachelard likes to describe himself very simply as a reader, not out of intellectual laziness or false modesty, but because of what happens when he reads: "is not the reader's imagination . . . revealed to be purely and simply the movement of quickly changing images?" (*AS* 288), and more strikingly, "it would seem that the reader is called upon to *continue* the writer's images, he is aware of being in a state of open imagination" (*TRR* 92). Reading poetic images brings us "the experience of *openness*, of *newness*" (*AS* 7), new images, new language, new possibilities in the world and in ourselves. We recognize a familiar pattern of thought. Yet Bachelard is not simply applying to poetry his own preconceived theories. What he brings to it is an attitude of mind, a willingness to accept and not reduce complexity, to take reading a poem seriously, as an aspect of our relationship with something other than ourselves.

What Bachelard reads is images, not ideas. In his first books, these are images of fire, water, air, and earth; later they are images of space—cellars and attics, shells, corners, the cosmos; and then in his last book, images of a candle flame. He reads material and dynamic images, neither perceptual nor rational, nor expressive of lived experience, images which are written, which are in and through language. As for how he reads, he was to begin with attracted by the psychoanalytical approach, because in 1938, when he wrote *La Psychanalyse du feu*, it was the only one sympathetic to the irrational material images he found in poetry. Yet Bachelard is always critical of psychoanalysis. He modifies and subverts Freud, and eventually, in his second series of books on poetry (1957–61), he rejects psychoanalysis, preferring phenomenology. He does so because psychoanalysis is reductive; it reduces images to the unconscious, the unconscious to lived experience, to infantile social experience in particular. In this way the diversity of poetic images is reduced to a strictly limited number of meanings, and no attention is paid to the fact that these are literary, written images. Bachelard modifies Freud by making the source of poetic images not the unconscious—for this would be to divide the subject as Bergson did, separating him from the object—but rather what he calls an "intermediate zone" on the threshold of consciousness and thought. Bachelard's material images, in which man and matter are conjoined, spring from "the zone of material reverie that precedes contemplation" (*ER* 6). His approach is never the diagnostic approach of the psychoanalytical critic; he is really interested not in the poet but in what the poet does to him: "literary images which are correctly dynamized will dynamize the reader" (*ER* 248). The poem acts on the reader not through some kind of unconscious communication but through language. In 1957, Bachelard turns from psychoanalysis to phenomenology precisely because this offers a better account of reading. *La Poétique de l'espace* (1957) and *La Poétique de la rêverie* (1960) are concerned first and foremost with reading, with the reader's consciousness of new language, of what he calls "the ecstasy of new images" (*PE* 1). However, he modifies Husserl as he did in his work on science, insisting on the dynamic relationship between subject and object, so that the reader's consciousness is changed by what he reads. The newness of the poet's images is linguistic: "through the newness of his images, the poet is always the source of language" (*PE* 4). It is this new language that changes the reader: "because of its newness, a poetic image sets in motion all linguistic activity . . . It becomes a new being in our own language, it expresses us by making us what it expresses . . . Here, expression creates being . . . Thus, the poetic image, an event of the *logos,* is for us personally innovatory" (*PE* 7).

For Bachelard, a poem is language, but this does not make him a crypto-structuralist. He never in fact reads a poem as a whole, as a structure of images, preferring to remain "on the level of separate images" (*PE* 9). Indeed, this does seem to be a weakness in the Bachelardian approach. "We shall not deal with the problem of the *composition* of a poem as a grouping of many images" (*PE* 8) because, as he explains, this involves complex cultural and historical factors and their influence on the poet, which he as a philosopher of science and not, one must remember, a professional literary scholar feels to be beyond his competence. Yet the limits of his approach are also its strengths: Bachelard presents a poem not as a cultural or linguistic phenomenon but as a personal experience. A poem is not something that confirms a preexisting body of knowledge, a theory or a hypothesis; it is "an explosive" (*AS* 285), a shattering and shaking of our foundations. When we read, we are in language, in language which is not our own. Thus far, he is in sympathy with the structuralists. This language—and here he differs—this autonomous "other" changes, renews, *opens* our own language.

"Poems are human realities" (*PE* 190): this, of course, sets Bachelard apart from the present generation of structuralists and poststructuralists. Yet he also insists that poems are written language, and long before Derrida and "grammatology," he makes this written language an experience not of "closure" but of "openness." For Bachelard, reading something that has been written is quite different from listening to someone speaking, for the simple reason that the spoken word imposes itself on us, requires our submission and our presence, whereas in the written word, read and slowly reread, "thoughts and dreams reverberate" (*AS* 285). The written word plays between the poles of subject and object; it interweaves and holds together ideas and dreams, the world and the poet, the text and the reader. In Bachelard's view, the language of poetry expresses at one and the same time both subject and object; it abolishes the frontiers of the internal and external worlds, making them reciprocal and interdependent. "Poems are human realities" because for Bachelard they exemplify our relations with the world, the imbrication of subject and object. We realize that in his work on poetry just as much as in his work on science, Bachelard is a subversive humanist.

"Man's being is an unfixed being. All expression unfixes him" (*PE* 193). This is surely important. Man is unfixed by language, not decentered. The question is not whether language is outside us or inside us. Bachelard discusses these notions of inside and outside in the penultimate chapter of *La Poétique de l'espace*, again anticipating Derrida in *De la grammatologie* (1967). Metaphysics, Bachelard declares, is bedeviled with this simple opposition, with this simple geometric intuition. It fails to see the complex human fact, and it is precisely this complex human fact that is the lesson of Bachelard's work. Mathematics and poetry, the two ways in which men use language most effectively, show how inadequate is this division between inside and outside. Language may well impose its rules on us—its differences are not of our choosing—but great men, mathematicians and poets, have shown how to shape it to new possibilities. Man is in language: how hard it is to escape this simple metaphor! Better perhaps, man is language: "all expression unfixes him." "Man is the being that lies half open" (*PE* 200), so that inside and outside flow together and are inseparable. This though is perhaps too simple, too static. Bachelard suggests another metaphor: "man's being is a spiral" (*PE* 193). Here, in the spiral, there is movement, and more important, no center. It is, in short, the image of Bachelard's subversive humanism.

This use of metaphor is characteristic in Bachelard, a matter not simply of his style but rather of his way of thinking. The polemical energy of his complex thought readily resolves itself into images. Others—René Poirier (1974:11), Georges Canguilhem (1957:5) for instance—have remarked on his style, on his refusal of what one might call traditional academic discourse. Yet to describe it as Canguilhem has done as "a rural philosophical style" is surely to miss the point, to ignore the intellectual tensions shaping it. My purpose in this book is to seek out these intellectual tensions. They are best seen, I believe, not in a general summary of ideas, for this is apt to disperse such tensions, but active in the text itself. It is for this reason that I have chosen to present Bachelard's ideas through the text, through translation, first and foremost. My own readings of the text should remain secondary. I offer them as signposts, as it were, to the implications of this subversive, allusive, highly metaphorical way of thinking and of writing. A study of Bachelard's texts reminds us, too, that he is a writer, a user and a maker of words. Philosophy, in his view, has as much to do with language as with ideas: "is not the function of the philosopher to deform the meaning of words, just sufficiently for him to draw the abstract from the concrete and allow thought to escape from things? Should he not, like the poet, 'give a purer meaning to the language of the tribe'? (Mallarmé) " (*II* 40). This practical concern with language is central to his work and, as this brief perspective has shown, to his conception of the human being. Bachelard deforms, subverts, unfixes language, which as he says is "there before thought" (*II* 40), and consequently we discover in his texts another tension, that between subject and object, whose old relationship of either submission or sovereignty he calls into question. The texts I have chosen represent the different aspects, the different phases of his thought, but they do more than just expound ideas; they embody them. Style, in Bachelard's case, is not simply the man, it is man using and used by language, man in and *against* a transcendent reality. Bachelard's texts both argue and exemplify his philosophy; they are the instruments of subversion, intellectual and linguistic, which the reader must confront.

Chapter 2
Polemics and Poetics: 1928–1932

> When you wish to know something, you immediately prepare for action, you alter the object and you alter the subject. Knowledge is one of the figures of change, the union of the other within the same.
> —*Essai sur la connaissance approchée*

The matrix of Bachelard's thought is, as we have seen, twentieth-century science, and his concern is to develop an epistemology adequate to the new science. This he does in four books published between 1928 and 1932: *Essai sur la connaissance approchée* (1928), his doctoral thesis of 1927; *Étude sur l'évolution d'un problème de physique: la propagation thermique dans les solides* (1928), his complementary thesis; *La Valeur inductive de la relativité* (1929); and *Le Pluralisme cohérent de la chimie moderne* (1932). The year 1932 also sees the publication of a very different book, so marking the end of this first stage in his thought. The book is *L'Intuition de l'instant*; the theme is time and consciousness. I have chosen three extracts to represent this first period, one from his *Essai sur la connaissance approchée* and two from *L'Intuition de l'instant*. Their juxtaposition points up what I have called the intellectual tensions which direct Bachelard's work from the beginning, namely the refutation of idealism alongside the retention of the subject.

How is it possible to reject idealism without also rejecting the subject? Bachelard's answer is adumbrated in the first of these passages, "Mind versus Reality: Bachelard's Theory of Approximate Knowledge," from chapter 1 of his *Essai*, "On Knowledge and Description." The broad lines of his argument are as follows. Bachelard rejects Kantian idealism, yet this rejection does not lead him to realism or to materialism. Reason remains of prime importance, while having lost its sovereignty. He in effect "decenters" reason, and with it, the knowing subject. Reason, he argues, is polemical. It is a function not of the structure of our mind, but rather of the struggle between mind and reality which he takes to be the motive power of knowledge. There are two poles of knowledge, as he puts it in this first extract, the knower and the known. Knowledge is like an alternating current oscillating between these poles (*ECA* 260), an important consequence of which is, in his view, the effect of knowledge on the knower. The knowing subject—and by implication his reason, his mind, and his consciousness—is *made* by his knowledge, and since scientific knowledge is always progressing and changing, dynamic and open, the subject will share these characteristics. Scientific knowledge is, Bachelard argues, both polemical and poetic.

This word *poetic* is used explicitly in its etymological sense of "making" in *La Poétique de la rêverie* (1960), when he wishes to show that the dreamer is made by his dreams (131). Characteristically, Bachelard subverts etymology to serve his own ends, to dramatize the interrelationship of subject and object: the dreamer is not simply the maker but rather is made by the dreams he makes. This notion of the "poetic" comes late in Bachelard's work, yet with hindsight we realize its presence from the very beginning, implicit even in his *Essai*. The knowing subject is made by his knowledge, but how? Put very simply (and in terms which are inevitably question-begging, but to which we shall soon return), the physicist, for instance, asks questions of reality, organizing it in certain ways in order to obtain an answer, though it is reality that arouses his curiosity in the first place, that prompts his questions. It is reality that by giving or withholding an answer arouses new curiosity and therefore changes the knowing subject. The subject is modified by the "not-self," the transcendent other that engages him in endless polemics. He both creates and is created by his knowledge of an external reality, in what Bachelard describes here as "the differential equation of epistemological movement." I have chosen to place the first stage of his thought under the rubric "Polemics and Poetics" in order to emphasize one particular consequence of his epistemology, namely its new, more complex understanding of the knowing subject.

If we turn now to the detail of this first extract, Bachelard's grounds for rejecting idealism are clear: it cannot account for "the continual movement, the continual progress of scientific knowledge," for "the fundamental incompleteness of knowledge" which he takes to be "an epistemological postulate." Later in his *Essai*, he will refer to epistemological discontinuity, to the "break" between twentieth-century science and its predecessors, to their "completely heterogeneous" concepts (270). This epistemological break is a fact; it proves the openness of scientific knowledge and disproves idealism, along with the notion of reason as an a priori deductive system, entirely necessary and entirely closed, independent of any reality. "Abstract ideas," Bachelard declares at the beginning of this book, must be "firmly and securely grounded in reality," adding that "description remains ... the aim of science." This somewhat surprising assertion introduces another intellectual tension in his thought, for he refuses realism while retaining reality.

Realism, with its notion of a given world, independent of the subject's mind and sense perceptions, is invalidated, he believes, by the facts not just of modern science but of ordinary knowledge. As he writes later in this extract, "something that is given has to be received," whether we are talking about subatomic particles, a child's experience of the world about him, or our efforts to know ourselves, to recover "the immediate data of consciousness." The reality that modern science describes is not a given reality, not a reality that lies waiting to be found, but rather, as Bachelard will argue throughout his work, a reality we construct. This may at first sound like reality-for-me, like idealism resurgent, yet this would be to misunderstand his thinking. He makes it clear that this constructed reality is not reality-for-the-subject, but rather reality-*against*-the subject. His conception of the interdependence of mind and reality, of subject and object, does not imply their unity, their convergence, their merging, but on the contrary, their "minimum opposition," as he calls it here, their difference.

Bachelard also affirms in this first extract that in modern science "reality's whole being is to be found in its resistance to knowledge," that scientific knowledge is "forced to twist and turn by rebellious matter." Characteristically again, he uses metaphor in order to resolve intellectual tensions, to take epistemology beyond rationalism and realism, beyond subject and object: this resistant reality is not that of realism, and this twisting, turning mind is not that of rationalism. The revolutions of twentieth-century science have dethroned both reason and reality, and forced them into alliance, a strange alliance based on conflict, on opposition and difference. Bachelard will, over the years, try a number of ways of describing this new relationship—"applied rationalism," "the philosophy of no," "the new scientific mind"—but his first attempt at description, which he in fact does not repeat, is perhaps the most interesting because of its richness of association and implication, his theory of "approximate knowledge."

In French, the title of Bachelard's thesis suggests rather more than its

English translation: *Essai sur la connaissance approchée* implies that we approach knowledge, that we do not arrive at it, once and for all. The knower is consequently always aware of a gap between what he knows and what there is to know, always aware therefore of difference, which he constantly strives to overcome only to find it again opposing him. Few commentators have discussed this notion of "approximate knowledge," and none, I suggest, has really understood why Bachelard used this term. Michel Vadée and Roch Smith are both interesting and helpful here, though Vadée thinks Bachelard was referring only to degrees of approximation, to the "second-order approximation" of modern science (1975: 46–50). Smith discusses this same aspect, although in more detail and more clearly, while he differs from the Marxist Vadée in stressing that this second-order approximation is not to a given, fixed reality, to a pole that will eventually be reached by dint of finer and finer approximations, for as he says, "certainty is in the process, not in the never-to-be-attained goal" (1982: 12). Both these interpretations are correct, yet both lose much of Bachelard's intentions. Vadée censures him for avoiding pragmatism, but without investigating why he does so. Smith, by stressing the "process," underplays Bachelard's own insistence on the "two paths" of science, on science as description.

I shall consider the question of pragmatism first. One of Bachelard's firmest polemics in his *Essai* is pursued against William James and pragmatism, to which, I would argue, "approximate knowledge" is in part a riposte. James's notion of usefulness or profitability as the criterion of truth he regards as quite untenable in modern science, in microphysics in particular where "success is relative and fragmentary" (267). Instead of success, Bachelard emphasizes failure and error: "error . . . is the driving force in knowledge" (249), "science is an enigma which is ever reborn" (155). We succeed in understanding something only to realize all that we yet fail to understand. Indeed, were it not so, the pursuit of knowledge would have been arrested long ago by man's first success. Thus, approximate knowledge is knowledge which, while seeking completeness, fails to be complete, which through verification and rectification seeks precision and certainty, only to discover its own error as it confronts a resistant, unknowable, inexhaustible reality. Pragmatism is, in Bachelard's opinion, a version of idealism (275), and as he declares in this first extract, it is undermined by "the undeniable existence of error . . . which obliges us to make do with approximations." Vadée both criticizes Bachelard's refusal of pragmatism and praises his notion of the new era of second-order approximation, without understanding that the latter leads to the former, that modern science with its new conception of reality, of matter, calls his own ideas, his dialectical materialism, into question.

The precise context of the *Essai* is important here: the revolution of quantum mechanics, the upheavals of the mid-1920s brought by the work of scientists like de Broglie and Schrödinger, Pauli, Born, Dirac, and Heisenberg. Bachelard's thesis was presented in the same year as Heisenberg's uncertainty principle, in 1927, and while direct influence is therefore unlikely—indeed there is no explicit reference to Heisenberg in the *Essai*—he seems peculiarly close to the implications of this principle. "Error," he insists, "is an inevitable and indeed essential part of knowledge" (280), and more specifically, "the phenomenon is absolutely inseparable from the conditions of its detection" (297). According to Heisenberg's principle, scientists cannot through experiment determine with exactitude both the position and the velocity of a subatomic particle, for when one is measured, the other is disturbed. That is to say, in observing and measuring, scientists disturb with their instruments what goes on naturally, so to speak; they introduce uncertainty into the question of the nature of things themselves, so that our knowledge of nature can only be approximate.[1] Exact knowledge is sought through mathematics, but again, since in microphysics this is the mathematics of probability, the notion of uncertainty, of approximation persists. How therefore can we continue to talk of "objective knowledge," of knowledge of "reality"?[2] Heisenberg's uncertainty principle in particular and microphysics in general challenged the old ideas of mind and reality, and of their separateness. Bachelard's theory of approximate knowledge is, I suggest, best understood in this context.

Roch Smith stresses that for Bachelard the pole of approximation—reality—is never attained. While this is correct, it also simplifies, for there is another aspect of approximation to be borne in mind. Bound up with this conception of approximate knowledge is the idea of approximating to irrational numbers, the problem that exercised early Greek mathematicians. Perhaps the most familiar example of an irrational is π which, as we have all learned, cannot be reduced to a whole or fractional number, remaining, as Bachelard himself says, arithmetically inexact, an example of approximate knowledge (*ECA* 188). Yet we use π in geometry, where it has a precise definition and where it gives precise results. Consequently, an irrational like π is knowable not in itself, but only in relation to the way knowledge is sought. It must also be remembered that irrational numbers are bounded by rationals, and that this bounding method is, in effect, approximation. Reason is in this way polarized by the irrational, which governs every mental inflection seeking to grasp it, but which always transcends the last approximation. Here then we see Bachelard's first conception of the new relationship between reason and reality. The irrational serves as a metaphor for reality, "unknown and inexhaustible," but at the same time imprinted, as it were, on reason, glimpsed in the twists and turns of approximation. Reason and reality are interdependent and reciprocal, reason constructing reality, and equally, reality constructing reason. Commentators have tended to distort Bachelard's thinking by emphasizing the constructive role of reason and neglecting that of reality. Reality may be unattainable, but it actively polarizes reason. It must not be forgotten that, as this first extract makes clear, "an inexhaustible reality" exerts "an attractive force," and so inflects "the mind's past."

Approximate knowledge is knowledge that is inexact, marked by failure and error. Yet it is more than this, for it is knowledge that "translates the richness of reality and its apparent contingency and irrationality" (183). This idea of the knower approaching reality is important because it limits the role of the subject, and also because it complicates that role, placing the nature of the subject in a new light. If the knowing subject can only approach reality, if in this second order of approximation knowledge is inexact and incomplete, then the subject is no longer sovereign. On the other hand, if this rich reality—"unknown and inexhaustible," resistant and rebellious, to which we can only approximate—is an irrational, then it has to be approached through reason, bounded by the rational subject. There is, in Bachelard's view, an unbridgeable gap, an ineradicable difference between subject and object which deprives each of them of sovereignty, and which paradoxically ensures their reciprocity.

This, Bachelard insists, is a "dynamic reciprocity," not a union, not a fusion, for "the contours of the object are modified by the knowledge that draws these contours, and the criteria of precise knowledge are dependent upon the object's order of magnitude, the stability of its appearance, upon, as it were, its order of existence" (250–51). Commentators have tended to stress the first aspect of this reciprocal relationship at the expense of the second, resulting in the overemphasis of the role of reason in Bachelard's thought, and in the charge of idealism. Yet Bachelard gives equal importance to the second aspect of the relationship between subject and object; he combats idealism by insisting that the object changes the subject. "The mind must mobilize in order to reflect the diversity and multiplicity of phenomena" (43); that is to say, the mind, in grappling with a rich, resistant reality acquires subtlety, richness, and diversity (92). Proof of this mutual dependence of mind and reality is microphysics, the "epistemological break" of post-Einsteinian science: nature may be finally unknowable, but it is nature that rules scientific knowledge, and consequently, in Bachelard's striking phrase, "the movement of description must bend in accordance with the curve of the universe" (283).

Extract I
Mind versus Reality: Bachelard's Theory of Approximate Knowledge

How are we to define knowledge? We know something when our description of that thing will allow us to find it on another occasion . . .

We would argue that knowledge obtained solely by deduction is limited; it can be no more than a well-ordered framework of thought as long as abstract ideas are not firmly and securely grounded in reality. Furthermore, the very fact that deduction proceeds by creating new abstractions requires that there be continual reference to the given world which, by definition, exists independently of logic. Where nature is concerned, we shall never see the day when the definitive, complete generalization has been made.

It would be incorrect, therefore, to say that real knowledge proceeds in one direction only. If we are to see knowledge at work, we must not hesitate to place it at the point at which it oscillates, where the "mathematical" and the "intuitive" mind converge.[1] Were we to give precedence to generalization rather than verification, we should be forgetting the hypothetical character of a generality whose only sanction is, in fact, its convenience or its clarity. When we begin to verify, we find that since total verification is impossible, generalization is somehow segmented and new problems are posed. Science progresses, then, along two paths.

If we are to resist the systematic approach to things that philosophers

---

1. Bachelard is referring to Pascal's *Pensées*, section 1, no. 1, in Brunschvicg's edition. Translation used: W. F. Trotter, *Thoughts* (New York: P. F. Collier, 1910), 7–9.

find so very attractive, we must give our initial description its full significance and never lose sight of the fact that description remains, when all is said and done, the aim of science. We must begin with description and we must in the end come back to description.

Description is often regarded as an inferior method, to be used only in the last resort, simply because in science more than anywhere else, we tend to confuse knowledge as it is communicated and knowledge as it is created. Consequently, the sign is given precedence over what it signifies; a good example is to be found in the physical sciences where mathematics is held to be a language that can easily be separated from its experimental basis, that can, as it were, think all by itself. If however we really wish to understand an experiment in physics, we have to translate the conclusions suggested by mathematics into the language of our personal experience. Our initial experiment was not analyzed by its more or less mathematical features, but only symbolized by them. How, in that case, could such conclusions ever rejoin reality? We see that description's job is never done, that sooner or later we shall have to return to concrete reality since, in our very first abstraction, the phenomenon itself has already been lost . . .

Whatever may be thought of this assimilation of scientific knowledge and description, it must be recognized that such an assimilation in no way prejudges the nature and destiny of thought. Description, in fact, requires a technique which will lead us back almost without our realizing it into the traditional paths of scientific progress. If we are to be able to use a description, that description must be ordered in relation to centers of interest which can be drawn closely together by a kind of high-speed shorthand. The organization of reference points will, whatever the principles on which it is based, inevitably lead to the kind of knowledge that tends to develop in the direction of maximum extension, relating similar qualities, and pretexts for identical action. This will, in due course, provide us with an overall view in which the conditions of generality are conditions of clarity; we shall then impose on reality not perhaps logical relations but at least rational ones, to use Cournot's definition of the term.[2] The given world will, in the end, be held in the grasp of a theory. Thus, in allowing ourselves to follow where description leads, we discover once again the human mind's fondness for systems.

2. Bachelard refers here to the applied mathematician and philosopher of science Antoine-Augustin Cournot (1801–77), who is best known for his work on the interpretation of the calculus of probability.

This autosynthesis seen here in description can take place in a number of different ways. Some privileged intuition, some tendency or impression may help toward it; indeed, it seems that anything at all can bring grist to the mill where the amalgamation of very different orders of experience is concerned. There are, we know, kinds of contact between the self and the not-self in which the unity of these two is achieved without delay, contacts such as those established in religious emotion or in the artist's vision, and apparently very different from any epistemological contact. Lichtenberger has pointed out the very special kind of intuition which unifies and sums up in the subject's emotions all the diversity of the poetic world.³ Here, then, a generality is created out of nothing, a generality that is quite plainly artificial and that has no foundation in reality. Goethe did not concern himself with the technical side of drawing, preferring to hold objects before his attention simply as opportunities for an emotional response. His sole aim was to grasp them as a whole, "in so far as they produced an effect." For the artist, then, we know something when our description of that thing will allow us to respond to it emotionally. Lyricism is, in our view, a coherent system in which the mind is especially agile and alive.

If we accept an ejective philosophy, we shall of course find it easy to make the subject the decisive factor in the convergence of self and not-self. We shall, in the end, arrive at a kind of truth that is essentially homogeneous and whose criterion is to be found in the agreement of thought with itself. There is, in fact, something even more important here. Never at any moment does knowledge lack a system, since reality is given only in so far as it accepts the mind's a priori categories. Here, according to this hypothesis, the elements of our description are chosen in such a way that this description will be closed in on itself. The method we use to find reference points turns into a method of construction, and knowledge is presented to us as something that is necessarily complete and final. This kind of knowledge is an arch which will stand only as long as all its constituent parts stand together. Its initial success is, in our view, very much a sign of weakness. We believe that idealism is, by definition, quite unable to follow and explain the continual movement, the continual progress of scientific knowledge. The systems in which idealism has put its trust cannot evolve slowly, through a kind of gradual deformation. We can overturn the forms of such systems for reasons of convenience, clarity, or rationality ... ,

---

3. Bachelard's footnote: Henri Lichtenberger, *Poésies lyriques de Goethe et de Schiller* (Paris: Hachette, 1892), 15. Note amended.

and here we are not forced to twist and turn by rebellious matter. Last, in idealism, knowledge will always be complete and yet closed to any extension. The only mobility such knowledge can ever know is the mobility of cataclysms.

We believe, too, that one of the strongest objections to idealism is the undeniable existence of error which, by its very nature, cannot be eliminated and which obliges us to make do with approximations.

Thus idealism is seen to be a fruitless and often specious working hypothesis when we are trying to understand the conditions in which epistemological progress is made. In actual fact, as Meyerson has proved, science usually postulates a reality.[4] *We believe that this reality, with all that is unknown and inexhaustible there, is eminently suited to give rise to an endless quest. Reality's whole being is to be found in its resistance to knowledge. We shall therefore regard the fundamental incompleteness of knowledge as an epistemological postulate.*

In saying this, we are not attempting to repeat the traditional modest disclaimer with which today's scientist limits the general scope of science. We must recognize these elements of uncertainty in the very life of science, present at every single moment and in every one of its endeavors. The act of knowledge is never a complete and final act. If it is performed with great ease, it is because it is developing on the plane of unreality. This unreality is the price that has to be paid for facility.

It seems at first sight that a study of approximate knowledge ought to repeat what Condillac tried to do, while seeking to avoid the artificial discontinuity of initiation.[5] *A new young brain should not be confronted with*

---

4. Émile Meyerson (1859–1933) was a philosopher of science whom Bachelard strongly criticized. This reference is unusual in its acceptance of Meyerson's ideas, perhaps acknowledging his importance as the only philosopher in his day to pay attention to recent developments in science. However, Meyerson was not interested in the results of science nor did he believe that science changes the way we think. Instead, he argued that the mind obeys its own rules, proceeding by deduction, reducing diversity to identity. Thus, Bachelard writes his book *La Valeur inductive de la relativité* (Paris: Vrin, 1929) against Meyerson's *La Déduction relativiste* (Paris: Payot, 1925), and frequently expresses his opposition to Meyerson's *Identité et réalité* (Paris: Alcan, 1908; republished 1912, 1926, 1932, 1951).

5. The empiricist philosopher Étienne Bonnot de Condillac (1715–80) is probably now best known in France for his story of a statue and a rose, by which he tried to show how all our faculties develop from sensation. In his *Traité des sensations* (1754), Condillac begins with a statue endowed with a sense of smell. A rose is then presented to this statue, and the statue is described as being conscious of itself as the smell of a rose, self-consciousness being nothing more than this sensation. The sniffing of the rose brings about the birth of

distinct and separate objects, but rather it should be set before a distant and indifferent panorama. Indeed, a sensation that hurts us has no lesson for us, nor can we learn anything from a sensation that prompts some adverse reaction. Knowledge must, from the very first, have in it an element of speculation. If sensation is to become representative, it must be gratuitous and it must be possible for us to inhibit its action and its consequences. Modern psychology has, in fact, drawn attention to the intellectual value of action. Action involves the will, and this is necessarily clear-cut, whole and undivided. All its endeavors bear the stamp of simplicity; it is by definition completely geometrical. Yet, once again, the lines that our action traces around things can establish only provisional and artificial reference points. Before action comes our reverie, miming a richer, more mobile world; there, the panorama of the given world imposes details upon us that ordinary, everyday action would habitually and quite safely ignore. Our actions are obviously far too crude to allow us to understand, in Bergson's use of the term, the given world in all its complexity and its delicacy of structure.[6] They are of value only in knowledge of a general and systematic kind. Now, we have no evidence that the world first commands our attention by means of its natural outline. Detail, picturesqueness, the unexpected and the accidental, all these awaken and amuse a young mind as it contemplates.

Yet how is a philosopher to conduct an enquiry here, in what is in fact the territory proper to the archeological study of childhood? He is like a historian who has no documents. Indeed, his research will always come up against a virtually insoluble question of principle. Even if we could hope to determine and relive the "immediate data" of consciousness, we do not see how we could restore the immediate mind. There is no logic that allows us to extrapolate its laws. And as for introspection, it is basically a culture, *for it can only objectify a memory through a personality that changes and grows stronger by virtue of its struggle to rediscover its origins.*

---

attention; if the rose is taken away, memory is born; if two roses are now offered to the statue, judgment arises, and so Condillac continues, showing how the mind is created by the external world it senses. Bachelard's use of the words "initiation" and "sensation" here suggests that it is this story of the statue that he has in mind.

6. Bachelard reminds us here of Bergson's conception of understanding and intelligence as being geared to action on the world, as relative to the practical requirements of everyday life, with the result that the given world is understood only in terms of its usefulness to us. This is a point of view that Bergson argues in all his books, and to which Bachelard is strongly opposed, as this paragraph shows.

We are wrong, moreover, when we confuse the primitive and the immediate. What is immediate for one person is not so for another. The given world is relative to our culture, and it is of necessity part and parcel of a construction. Were it entirely without form, pure and irremediable chaos, then reflection would be quite unable to come to grips with this given world. Yet conversely, were the mind without any categories or any habits, then the function "given" would, if we consider the real meaning of the term, be completely nonsensical. Something that is given has to be received. We shall never manage to separate the order of the given world from the method we use to describe it, nor shall we ever be able to merge these two completely. These two terms represent, in our view, the minimum opposition of mind and reality, and between them there are constant reactions, reactions that give rise to reciprocal resonance. At every single moment, therefore, a renewed given world is offered to our mind . . .

These, then, are the reasons that lead us to consider knowledge as it flows along, far from its origins in the senses, at the point at which it has become intimately involved with reflection. Here alone will knowledge possess its full meaning. The source of a river is simply a geographical point: the river's power and momentum do not lie there.

Knowledge that is forever moving onward is therefore a kind of continuous creation; the old explains the new and assimilates it, and vice versa, the new strengthens the old and reorganizes it . . . The mind that knows must of necessity have a past. The past, the antecedent provide, as we shall later demonstrate from different points of view, the tools with which we can arrive at explanations. Besides, is not the living mind distinguished from the inert object by the riches which lie ever at our disposal, which memory uses in accordance with the situations that arise, so that our actions may be adjusted to fit new circumstances? It is this inflection of the mind's past as a result of the attractive force exerted by an inexhaustible reality that constitutes the dynamic element of knowledge. If we relive its impulse, we can, as it were, formulate the differential equation of epistemological movement. Rectification is not, in our view, just a matter of turning back to an unhappy experience that we can correct if we pay it closer and better informed attention. Rectification is, in fact, the fundamental principle which upholds and directs knowledge, and which never ceases to thrust knowledge forward to make new conquests.

Essai sur la connaissance approchée (1928; Paris: Vrin, 1969), 9–16.

L'Intuition de l'instant seems at first sight to differ sharply from Bachelard's *Essai* with its theory of approximate knowledge. Its theme, as the two following extracts will show, is time and consciousness, and therefore, by implication, the nature of the subject. Though the emphasis here is clearly on the subject, this does not mean a betrayal of Bachelard's epistemology, as Roch Smith seems to be suggesting, nor indeed as Michel Vadée inevitably claims, is it confirmation of Bachelard's idealism. According to Vadée, Bachelard is a Bergsonian, conceiving the subject as isolated from the world. Bachelard on the contrary defines himself *against* Bergson, precisely because this separation of the subject from the object ignores what he calls in the first of these two extracts "the lessons of science." Chief among these lessons, as we have seen, is the "dynamic reciprocity" of reason and reality, the "inflection of the mind's past as a result of the attractive force exerted by an inexhaustible reality." This last phrase, and indeed the concluding paragraph of the preceding extract as a whole, indicates that already in the *Essai*, Bachelard conceives scientific thinking as temporal, as the interaction of past and future, so preparing the way for his argument against Bergson in *L'Intuition de l'instant*.

The notion of consciousness is also introduced very early in the *Essai*, as the previous extract has shown, when a child's consciousness of the world is briefly discussed. What is most striking here is that Bachelard equates consciousness with knowledge, with thinking about or rather *against* the world. He is clearly opposed to Bergson, implicitly rejecting both *Matière et mémoire*, with its insistence on the priority of the body and its response to "our multiple needs," and also his *Essai sur les données immédiates de la conscience* (*Time and Free Will*), in particular its conception of the noninterference of introspection in "pure" consciousness. For Bachelard, a child's consciousness is the response of his mind to the multiplicity of the world, "rich" and "mobile," "complex and delicate," a world which not only awakens the mind but ensures that it too is rich and ever-changing. "Consciousness is . . . a function of the mobility and therefore of the number of points of view," he argues later in his *Essai*; "in order to constitute the self, as indeed the object, we need an epistemological plurality" (259). If "rêverie" (surely a fascinating variation on approximate knowledge) "mimes" a world of "detail," of "the accidental," there can be no "immediate mind" but only introspection mediated by the object of its scrutiny. Applying "the lessons of science" to knowledge in general in his *Essai*, Bachelard rejects Bergson's conception of pure consciousness, of pure duration. "We have to accept science as a whole," he declares in the first of the two following extracts from *L'Intuition de l'instant*, and this includes any repercussions it may have on our understanding of ourselves.

L'Intuition de l'instant is not a departure from science but rather a reaffirmation of the need to take the fullest possible account of the consequences of its developments in the twentieth century. While the book is ostensibly inspired by his friend and colleague Gaston Roupnel's *Siloë*, it remains, as Jacques Gagey has so convincingly shown, untouched in its essentials by Roupnel's thinking. Bachelard himself fears he may falsify this by what he calls his own "arabesques" (8), and it is indeed true that he dances to his own music. Gagey argues that Roupnel (a colleague at Dijon since 1927 when Bachelard began to teach part-time at the university) does not so much influence Bachelard as "convert him to himself" (1969: 48), to "an awareness of the poetic dimension of existence" (56), to "a desire for poetry" (81), an idea echoed by Roch Smith in his turn (1982: 56–57). Yet this is surely to look too far ahead, to polarize Bachelard's early books by his later work on poetry. Bachelard's "conversion to himself" is also—and less controversially—evident if we restrict ourselves to this first period of his work, to the years 1928–32, since *L'Intuition de l'instant* develops ideas of consciousness and time already present, though briefly, in his *Essai*. More important, Einstein's theory of relativity, the avowed instrument of Bachelard's conversion, was the subject of a previous book, *La Valeur inductive de la relativité*, published in 1929, with regard to which *L'Intuition de l'instant* should therefore be considered.

It is in the introduction to *La Valeur inductive de la relativité* that Bachelard first expounds at length and with a degree of emotion his idea of the epistemological break, of the *newness* of relativity theory, which was neither derived from experiment nor deduced from Newtonian physics. The novelty and boldness of his approach emerge as he takes on the then influential philosopher Émile Meyerson, whose book *La Déduction relativiste* (1925) is plainly his target. One important aspect of this new theory is, Bachelard argues, its break with habits of thought where absolutes are concerned, that is to say, with our tendency to make either the subject or the object, either mind or reality absolute, each separate and self-contained. This he discusses in chapter 3. Relativity forbids such a view; it "uncrowns" these absolutes, teaching us "to grasp relations independently of the terms related" (*VIR* 97–99). "In the beginning are relations" (*VIR* 210), and this phrase is repeated in an article a year or so later, "Noumène et microphysique" (*Ét.*, 19), where he adds, "this is why mathematics rules reality." It is important to look at this last phrase in its context. Mathematics "rules" reality not as a sovereign with absolute power, but because of its ability to coordinate, relate, and reveal variables, discontinuities, diversity, because as he puts it in another contemporary article, it is equipped "to explore the riches of reality" (*Eng. rat.*, 116). Relativity theory, like all twentieth-century physics and chemistry, shows reality to be mathematical, "formed by amassing relations," as he writes towards the end of *La Valeur inductive de la relativité*, a reality which is not "found" but "conquered" (241). Consequently, and it is this that is perhaps most important for Bachelard, we are no longer in a lived but in a *thought* world.

Bergson, in *Durée et simultanéité* (1922), discusses Einstein's theory and quite explicitly equates reality with what is observed, perceived, and *lived*. "Real" time is for him "lived time"; it has unity, and is not in any way affected by relativity theory. To Bachelard, this "commonsense" attitude is not only untenable but psychologically damaging. Nevertheless, as he says at the beginning of "Noumène et microphysique," this same harmful attitude dominates school science teaching, with its emphasis on experiment, on "looking and learning" (*Ét.*, 12). Formerly a schoolmaster himself—he taught physics and chemistry at the *collège* in Bar-sur-Aube between 1919 and 1930—he realizes the attractions of such an approach, and yet at the same time, as a philosopher of modern science, he knows its limits. It deprives children of the science of their own century, and even more important, of "the psychologically dynamic and inventive value" of mathematical realism (*Ét.*, 17). *L'Intuition de l'instant* should be read against this background, in terms of first the constructed, *thought* reality that is for Bachelard a scientific fact, and second its psychological aspect, its effect on our understanding of consciousness.

"We do not *think* real time," Bergson declares in *L'Évolution créatrice*, "we live it" (1962: 46 [1919: 49]), and proves this with the well-known example of the sugar lump: the time I wait for a lump of sugar to dissolve is "my impatience," my "duration"; it is therefore "lived" not "thought" time (1962: 9–10 [1919: 10–11]). It is significant that Bachelard should refer to this rather than to an example of duration from one of Bergson's earlier works, for it is in *L'Évolution créatrice* that thought, and with it science, is most insistently downgraded. At the same time, Bergson seeks to ensure the possibility of objective experience through the coincidence of "my duration" with that of the universe (1962: 11 [1919: 11–12]), and of consciousness with "organic evolution" (1962: 27 [1919: 29]), by positing a "life force" (1962: 88–98 [1919: 92–102]), a "current of consciousness" in all matter (1962: 182 [1919: 191]). This curious account of objectivity, together with the disparaging of thought, must have been anathema to Bachelard. He is convinced that relativity not only invalidates Bergson's conception of absolute duration but undermines the whole basis of his argument. Initially, in Bergson's *Essai sur les données immédiates de la conscience* (*Time and Free Will*), the idea of duration, of immediate consciousness, was developed in order to prove human freedom, in opposition to thought, to the "superficial self," held to be determined by the requirements of social and practical life. This fundamental distinction between thought and "immediate consciousness" no longer holds, for in modern science, Bachelard argues, thought is not determined by the need for practical action. Instead, it is characterized by contingency, by the arbitrary and the possible, that is to say by freedom. There are no longer any grounds for dividing the subject as Bergson did, and as a result, consciousness and thought, and with them the subject and the object, must be brought together.

If for the sake of argument the "lessons of science" are ignored, Bachelard believes that Bergson's ideas of time and consciousness are still invalidated by his own preferred tool, introspection. This first extract here, "Duration, Instants, and the Absolute: Bergson or Einstein?" from chapter 1, "The Instant," shows Bachelard following Bergson's instructions, detaching his consciousness from all objects, and trying to reach "pure duration" where there is "succession without distinction," "mutual penetration, solidarity" (Bergson 1961: 74–75 [1910: 101]). However, Bachelard's introspection yields very different results: heterogeneity, multiplicity, instants. Most important of all, he discovers the impossibility of isolating consciousness from the world, from the *aggression* of the world, "the continual onslaught of all that is accidental and new." Bergson uses music as a simile for duration, as a model of consciousness: the notes of a melody may follow one another, but we hear them as a whole, one note merging into another, inseparable, changing qualities. Bachelard argues against this view of music in the second of these extracts, "On Consciousness and Time," from chapter 3, "The Idea of Progress and the Intuition of Discontinuous Time". There are few references to music in his work, but this description of music, with its insistence on diversity, on difference even in "the purest sound," appears more accurate than Bergson's. It confirms moreover his conception of consciousness as a multiplicity of instants.

There is a problem here as Vadée and Smith have pointed out, since this juxtaposition of the "lessons" of science and those of personal introspection leads to ambiguity, if not contradiction. On the one hand, the Einsteinian instant is held to be an absolute, so that logically, as Vadée says, there should be only one, unique instant, while on the other, the contrary view is presented, that of reality as a multiplicity of instants. Vadée regards this "metaphysical *contradiction*" giving precedence to the unique instant and making it contain multiplicity as further evidence of Bachelard's idealism (1975: 124). Smith sees Bachelard's solution to the problem as not philosophical but aesthetic, the metaphor of music embracing both the reality of the instant, i.e., the note, and the reality of multiplicity, i.e., rhythm (1982: 58–61). Attractive as this is, it neglects any reference to the fact that this metaphor is used against Bergson, that Bachelard's intention is to invalidate Bergsonian duration. Indeed, as this second extract here shows, what Bachelard emphasizes is the diversity of each single note, none of which does he appear to regard as ever being unique or absolute. A careful reading of both these extracts suggests that Bachelard's order of priorities is in fact the reverse of that argued by Vadée, and the book as a whole bears this out, with its many references to multiplicity, diversity, variety, heterogeneity, and progress, to instants as "new," "creative," "distinct because they are fruitful" (86). Nonetheless, the idea of the instant as an absolute remains and although inconvenient, it cannot be ignored. Both aspects of the problem can be reconciled if it is approached in terms of Bachelard's argument against Bergson. The absolute, Einsteinian instant is invoked as a scientific fact profoundly damaging to Bergson's theories, just as the multiplicity of instants of consciousness is regarded as an equally damaging experiential fact. Thus, Bachelard is seen to be attacking Bergson from all sides. The notion of the absolute instant is best understood as a weapon, ad hoc and expedient, rather than as a fundamental, cherished idea. As Hervé Barreau has pointed out (1974: 335–39), the idea of the instant as an absolute in relativity theory is scientifically questionable, and will be swiftly abandoned by Bachelard.

More important here is the conception of consciousness as a multiplicity of instants. In this first passage, Bachelard explains it in terms of "the continual onslaught of all that is accidental and new," suggesting a polemical relationship between the subject and the nonsubject, the "other." This idea is reinforced and developed when, in the second passage, he states that consciousness is enriched if we "multiply our thoughts," if we maintain it as "always active, never passive," if instants are always "being used." The instant for Bachelard is therefore an instant of thought *against* the world, knowledge coming to us from "the world's attack" (36). "The world brings us knowledge," he continues, "and it is in a fruitful instant that our attentive consciousness will be enriched with objective knowledge." Underlying Bachelard's ontology is, we realize, his epistemology. His *Essai sur la connaissance approchée* with its conception of the interdependence of reason and reality leads, if indirectly and unexpectedly, to a theory of time and consciousness.

Extract II
Duration, Instants, and the Absolute: Bergson or Einstein?

Bergson's idea of duration found further support in the many proofs of its objectivity that he was able to marshal. Certainly, Bergson asked us to feel duration within us, in our personal experience of our own innermost self. He went further than this, however. Bergson showed us objectively that we are all part of one single life force, all swept along by the same tide. Should boredom or impatience make an hour last longer, or happiness shorten our day, then impersonal life, the life of other people, will remind us of the real nature of Duration. We have only to watch a simple experiment, a lump of sugar dissolving in a glass of water, and we shall realize that there is indeed an objective, absolute duration that corresponds to the duration we ourselves feel.[1] This being so, it was argued that Bergsonian theory here reentered the realm of measurement, while still preserving all that we know through the intuition of our innermost self. Our soul is in immediate contact with the temporal quality of being and with the essence of its becoming; yet the objectivity of becoming belongs to the realm of time as a quantity, notwithstanding our indirect and limited knowledge of this. It seemed, then, that both discursive proof and the evidence of intuition safeguarded the primitive character of Duration.

---

1. Bachelard is referring to Bergson's *L'Évolution créatrice* (1907; Paris: Presses Universitaires de France, 1962); see *Creative Evolution*, authorized translation by Arthur Mitchell (London: Macmillan, 1919).

Polemics and Poetics

Let us now explain why our own confidence in Bergson's argument was shaken.

We were aroused from our dogmatic slumbers by Einstein's criticism of objective duration.

It was very soon clear to us that Einstein destroyed the absoluteness of that which has duration, while maintaining, as we shall see, the absolute character of what is, that is to say, the absolute character of the instant.

It is the *interval* of time, the "length" of time that becomes relative in Einstein's theory. Length of time is shown to be relative to the method of measuring it. It is said that, were we to take a trip in space at a high velocity, we should find on our return that the earth was a few centuries older whilst, according to the clock we took with us, only a few hours would have passed. A much shorter journey would be required to adjust the time it takes for a sugar lump to dissolve in a glass of water, a time which Bergson postulates as fixed and necessary, so that it becomes synchronous with our impatience as we watch and wait.

We are not just playing pointless mathematical games here, and we would emphasize this straightaway. The relativity of intervals of time in systems which are in motion must now be regarded as a scientific fact. Were it considered legitimate to reject the lessons of science in this respect, we should have to be given the right to entertain doubts concerning the intervention of physical circumstances in the sugar-dissolving experiment, and doubts, too, about the active interference of time with experimental variables. For instance, we all accept that temperature is involved in this experiment with the sugar lump. Very well then, let us say that for the modern scientist this experiment also involves the relativity of time. We have to accept science as a whole and not try to limit it to just some of its aspects, beyond which we refuse to go.

In this way, then, relativity brought the sudden ruin of everything to do with the external proof of one unique Duration, that fundamental principle ordering all events. The Metaphysician had to retreat into his own local time, shutting himself into the duration deep within him. The world offered no immediate guarantee that our own individual duration, the duration that each of us lives within our own private consciousness, would ever converge with the duration lived by another person.

We now come to something to which it is well worth paying close attention. The instant, clearly and accurately defined, remains in Einstein's theory an absolute. We can give this value to the instant simply by considering it in its synthetic state, as a point in space-time. In other words, we must accept that being is a synthesis comprising both space and time. It lies at the point of concurrence of place and present time: *hic et nunc;* not here and tomorrow, not there and today. In these two formulae, the point would expand along the axis of duration or along an axis of space; in each of them, there is something that makes a precise synthesis impossible, so that they would both lead to an entirely relative study of space and duration. Yet the moment we agree to weld and fuse together these two adverbs *hic et nunc,* the verb "to be" will at last receive its full force: it is the absolute.

In this very place and at this very moment; here simultaneity is unmistakable, it is evident and precise; here, succession is ordered without any hesitation or any obscurity. One consequence of Einstein's theory is that we can no longer claim that the simultaneity of two events localized at different points in space is self-evident. For this simultaneity to be established, we should need an experiment in which we could be sure, beyond a shadow of doubt, that the stationary ether exists. Michelson's failure put an end to all hopes of ever carrying out such an experiment. We must therefore arrive at an indirect way of defining simultaneity in different places, and, as a result of this, we shall have to adjust our measurement of the duration separating different instants in conformity with this new and always relative definition of simultaneity. There can be no true concomitance without coincidence.

We return, then, from our foray into the realm of the phenomenon, convinced that duration accumulates in a purely artificial way, in a climate of preexisting conventions and preliminary definitions, convinced also that its unity derives from nothing other than the generality and the sluggishness of our own investigation. Conversely, the instant is shown to be capable of precision and objectivity, and we sense here, in the instant, the character of something fixed and absolute . . .

When we still accepted Bergson's idea of duration, we set out to study it by trying very hard to purify and consequently impoverish duration as it is given to us. Yet our efforts would always encounter the same obstacle, for we never managed to overcome the lavish heterogeneity of duration. We laid the blame on our own inept meditation and on our failure to detach ourselves from the continual onslaught of all that is accidental and new. Never at any time did we manage to lose ourselves so completely that we then truly found ourselves, nor did we ever meet with success in our efforts to reach and then follow something in us that is uniform and flowing, where duration slowly sets before us a life story in which nothing is lived, a happening in which nothing happens. We would have preferred it if becoming were like a flight in a cloudless sky, disturbing nothing and hindered by nothing, soaring higher and ever higher into emptiness; in short, we would have preferred to find becoming in all its solitude, in all its purity and simplicity. Many a time have we sought to discover elements in becoming which would be as coherent and as clear as those Spinoza found in his meditations on being.

We remained, even so, quite incapable of finding these endless, unbroken lines within us, the simple sweeping lines traced by the life force in the picture it draws of becoming. In due course, as one might expect, we tried to find this homogeneous character of duration by confining our study to smaller and smaller fragments. Yet we were still dogged by failure. There was far more to duration than just having duration, for duration was truly alive! However small the fragment under consideration, we had only to examine it microscopically to see in it a multiplicity of events; always it was the embroidery that we saw, never the fabric, always the shadows and reflections mirrored by a restless river, never its deep, pellucid waters. Duration, like substance, can offer us only phantoms, only shadowy things. Indeed, duration and substance endlessly enact the fable of the deceiver deceived, the one deceiving the other only to be in its turn inevitably deceived, for becoming is the phenomenon of substance and substance is the phenomenon of becoming.

Should we not agree, then, that it is metaphysically more prudent to equate time with what is accidental, which would mean, in fact, that time is to be equated with its phenomenon? Time can be observed solely in the instant; duration, as we shall see, can be experienced solely in the instant. Duration is a cloud of instants, or better still, it is a group of points which are drawn closely together by the *phenomenon* of perspective.[2]

L'Intuition de l'instant: étude sur la "Siloë" de Gaston Roupnel (1932; Paris: Gonthier, 1966), 28–33.

---

2. Bachelard's footnote: it has already been said by Guyau, though from a more psychological point of view than ours, it is true, that "the idea of time . . . is in the end the result of a kind of perspective." (*La Genèse de l'idée du temps*. Preface). My translation. His reference is to p. 33 in Jean-Marie Guyau's book (Paris: Alcan, 1890); not translated.

Extract III

On Consciousness and Time

Bergson wishes to simplify time as we first experience it, and in order to do this, he takes a piece of music as his starting point. However, instead of stressing that music has meaning only because of the diversity of its sounds, and instead of recognizing that each sound has itself a diverse life, Bergson tries to show that, by eliminating this diversity not just of different sounds but even within a single sound, we shall in the end reach uniformity. In other words, if we remove from a sound everything in it that can be perceived by our senses, we should then discover the uniformity of fundamental time. This, in our view, will only lead us to the uniformity of nothingness. If we examine a sound which, objectively, is as unified as possible, we shall see that this unified sound is not uniform subjectively. It is quite impossible to preserve any kind of synchronism between the rhythm of the stimulus and the rhythm of our sensation. Even the most cursory acquaintance with music will make us realize that our perception of sound is not a simple process of addition, and that since vibrations do not occupy the same space, they cannot each have an identical role. We have proof of this in that a sound which is prolonged without any variation becomes truly agonizing, as Octave Mirbeau has so perceptively observed.[1] The very same case against uniformity can be made in every sphere, for repetition, pure and simple, has similar effects in both the organic and the inorganic world. When repetition is too uniform, it spells destruction for the very hardest material which, when subjected to certain monotonous rhythms, will always finally disintegrate. This being so, we surely cannot develop our discussion of the psychology of acoustic sensation by echoing Bergson when he talks about "the continuation of what precedes in what follows" and "uninterrupted transition, multiplied yet without diversity," and "succession without separation"; we have only to prolong the purest sound for its character to change.[2] We can, moreover, find evidence of this

---

1. Octave Mirbeau (1848–1917) was a prolific novelist, dramatist, and critic, who dealt with contemporary problems, but is now largely forgotten. His best play, *Les Affaires sont les affaires*, first performed in 1903, and on the theme of big business, is the only one still read (Paris: Fasquelle, 1924). Bachelard gives no indication of what he refers to here. This reference is further evidence of the breadth of Bachelard's reading.
2. Bachelard is referring to Bergson's *Durée et simultanéité* (Paris: Presses Universitaires de

without having to take the rather extreme example of a sound which becomes painful through its prolongation, for if we allow a sound its full musical value, we have to recognize the fact that, when it is prolonged in a moderate way, without exaggeration, the sound is renewed and it sings. The closer the attention we pay to an apparently uniform sensation, the more it will diversify. Anyone who can entertain the idea of discovering a kind of meditation which would simplify sense data is really and truly the victim of abstraction. Sensation is variety, and it is memory alone that confers uniformity. There is always, then, this same difference of method that separates us from Bergson; Bergson takes time which is full of events, taking it moreover at the point where there is consciousness of events, and then he gradually obliterates these events, or rather the consciousness of these events; he believes that he would now reach time without any events, and indeed consciousness of pure duration. Our own approach is the very opposite of Bergson's here, for we can only feel time if we multiply instants of consciousness. Should indolence lull our meditation, it is not unlikely that there will still be sufficient instants enriched by the life of the senses and of the flesh for us to remain vaguely aware that we have duration; if, however, we wish to sharpen and explore this awareness, there is, in our experience, only one way to do so and that is by multiplying our thoughts. For us, consciousness of time is always consciousness that instants *are being used, it is always active, never passive; in short, consciousness of our own duration is consciousness of* progress *in our innermost being, whether this progress be real or counterfeit, or even simply dreamed. Once complexity has been organized into this kind of progress, it will be much clearer and simpler, and similarly rhythms which are properly renewed will be more coherent than pure and simple repetition. Furthermore, if this conscious, careful construction means that we eventually arrive at uniformity in our meditation, it seems to us that here we have a new conquest, for this uniformity is found when there is an ordering of creative instants, in for example one of those general, fertile thoughts which embrace and command a thousand ordered thoughts. Duration is richness, then, and it cannot be found by abstraction. We weave the fabric of duration when we place instants in sequence, one after the other, so that no instant ever touches another, and all these instants are concrete and rich in their conscious newness, a*

France, 1922), 42. Despite the quotation marks, he does not give his source, and in fact misquotes! For this reason, I give my own translation here. See *Duration and Simultaneity*, trans. Leon Jacobson (Indianapolis: Bobbs-Merrill, 1965).

newness which is always carefully controlled, never excessive. Duration has coherence if and only if we can coordinate some kind of method of enrichment. We cannot speak of pure and simple uniformity, unless it is with reference to a world of abstractions, or to a description of nothingness. It is along the path of richness, not of simplicity, that we must journey if we would reach the limits of experience.

The only truly uniform duration is, we would argue, a duration which is uniformly varied, a duration which is, in effect, progressive.

L'Intuition de l'instant: étude sur la "Siloë" de Gaston Roupnel (1932; Paris: Gonthier, 1966), 86–89.

L'Intuition de l'instant reveals what Roch Smith has called "an unprecedented reliance on subjective experience" (1982: 57). It must also be stressed that this is the subjective experience of thought. The "limits of experience" referred to in this second extract and toward which Bachelard strives "along the path of richness" are the limits of thought, of approximate knowledge. Indeed, he describes the "instant of nascent knowledge" very strikingly in his introduction to *L'Intuition de l'instant*: it is "kept active" by "consciousness of the irrational"; it is "the truly synthetic moment" where failure and success are one (6). Over and above all else, *L'Intuition de l'instant* shows how Bachelard's epistemology has become charged with value, partly perhaps as a result of his friendship with Roupnel and of reading his *Siloë*. The blind man in Bachelard's adaptation of the story of the pool of Siloam is he who refuses to think, who isolates himself from "the world's attack" in the Bergsonian consciousness. The restorer of sight is not just reason but the "irrational" (5–6), which in a curious and idiosyncratic amalgam of mathematics and metaphysics becomes the "divine redeemer" (95). Approximate knowledge opens our eyes, so to speak, to a new, polemical relationship between reason and reality, and consequently to a new view of ourselves. It spells the end of our tranquil sovereignty and equally of our tranquil submission, making our relationship with the "not-self" mutually aggressive. What emerges from Bachelard's first books is the exceptional fruitfulness of this polemic. His next concern will be both to sustain and promote its creativity.

Chapter 3

The New Scientific Mind: 1934

> The metaphysical gulf between the mind and the external world is and will remain unbridgeable for immediate, intuitive metaphysics. It seems less wide to the discursive metaphysics that seeks to follow developments in science... The philosopher who attends the lessons of quantum theory—the *schola quantorum*—will agree to think all reality in its mathematical organization. In other words, he will get used to measuring reality, metaphysically speaking, in terms of possibility, to what is in fact the very opposite of realism.
> —*Le Nouvel Esprit scientifique*

In 1930, Bachelard became professor of philosophy at the University of Dijon, remaining there until his appointment to the chair in the history and philosophy of science at the Sorbonne in 1940. These ten years are productive ones: ten books, and some twenty-four articles, book reviews, and conference papers are published, the latter involving travel in France and abroad, to the Hague, Ghent, Amersfoort, Cracow, and Prague. They are the years of Bachelard's maturity (he was forty-six when he went to Dijon), years marked by considerable intellectual confidence as he moves beyond his immediate professional sphere, into ontology with *L'Intuition de l'instant* in 1932 and *La Dialectique de la durée* in 1936, and then, toward the end of the decade, into poetry, publishing *La Psychanalyse du feu* in 1938 and *Lautréamont* in 1939. This diversity of interest is reflected in the subject matter of his articles and reviews: perception (1934), language (1934), imagination (1937), Buber's *I and Thou* (1938), Marie Delcourt's *Stérilités mystérieuses dans l'antiquité* (1939). Where his work on science is concerned, that too is varied, combining history and epistemology in *Le Pluralisme cohérent de la chimie moderne* (1932) and *Les Intuitions atomistiques* (1933), psychology and epistemology in *Le Nouvel Esprit scientifique*

(1934), eventually outlining in *La Formation de l'esprit scientifique* (1938) what he calls a "psychoanalysis of objective knowledge." None of his publications is a routine, predictable discussion, even when the topic is mainstream. As his chosen titles suggest, his approach is always idiosyncratic: one article is on "the noumenon and microphysics," another is on "surrationalism," while the last book of the decade discusses "the philosophy of no." It is almost as if Bachelard cultivates novelty.

Yet if he does, it is neither for its own sake nor because of current fashion, for Bachelard never runs with the mob. These new directions are not followed at random. On the contrary, as even the briefest survey of the decade shows, there is a pattern here, a growing and more general interest in the mind, the consciousness, the psychology of what he calls *homo aleator*. *Le Nouvel Esprit scientifique*, where this phrase occurs, is the pivot of this decade. Published in 1934, when Bachelard was fifty years old, its subject is both the epistemology of modern science and quite explicitly the "psychology of the scientific mind" (48, below). The two extracts chosen from this book illustrate and in some measure explain this dual interest. The first is from the introduction, "The Essential Complexity of the Philosophy of Science"; the second is a passage from the last chapter, "Non-Cartesian Epistemology." I shall center my discussion on three statements, two from the first extract, one from the second. First, "science in fact creates philosophy. The philosopher must therefore inflect his language so that it can express the supple, mobile character of contemporary thought." Second, "modern science is founded upon the *project,* above the *subject* and beyond the immediate *object*. In scientific thought, whenever a subject thinks about an object, his reflection is in the form of a project." Third, "it is, moreover, when the subject thinks about an object that this subject is most likely to acquire depth."

"Science in fact creates philosophy." This simple phrase tends to hold the reader's attention, summing up as it does the theme of all Bachelard's epistemological work. It is, however, what follows that is important, because it defines not only philosophy but the philosopher: "the philosopher must therefore inflect his language so that it can express the supple, mobile character of contemporary thought." The philosopher is a user of language, and the problems he faces are as much linguistic as conceptual because of the inadequacy of ordinary language to modern science. Why persist, then, in an enterprise that is bound to fail? Why not remain in the safe harbor of mathematics? The fact that Bachelard does persist

The New Scientific Mind

suggests the value he ascribes to modern science for everyone, scientist, philosopher, and "common man" alike. What he values is not primarily concepts, but "the supple, mobile character of contemporary thought," so that if he inflects his language, as he puts it, it is to communicate to his readers the quality, or what we might call the "feel," of the "new scientific mind," to place us, as it were, inside the scientist's head. He goes on to refer three times in this first extract to the "psychology of the scientific mind" and once to the "psychology of the mathematician," phrases he has not used before, and indicative therefore of a change of emphasis and a new direction. What constitutes this "psychology" is familiar to us from his previous work, "the synthesis of metaphysical contradictions," the dynamic interdependence of reason and reality, reason constructing reality through induction, dialectics, polemics. Psychology would seem to be epistemology seen from another angle, in terms of the subject's experience of scientific thinking. So when Bachelard inflects his language, he has two aims in view: to help his reader understand what is happening in science, and also how this affects him, not materially but in his status as a subject.

"Science . . . creates philosophy." Modern science, as Bachelard argues throughout this first extract, requires philosophers to break with the old definitions of reason and reality as alternative and absolute. This cannot be done simply by redefining words, for if an old word continues to be used, its old meaning will remain. The philosopher must therefore adapt familiar words to make them express new ideas adequately, to help his readers understand how the new science differs from the old. With "realization," for example, Bachelard adapts "reality" in order to stress that in the new science reality is constructed by reason. "Technical realism" and "phenomeno-technique" are invented to make the same point. The two meanings of "applied" are exploited to show not just the practical, experimental aspect of science but that reason and reality are "overlaid," reciprocal. He likes to bring together words that are similar in some way, through assonance, through common syllables, using this resemblance to dramatize the difference between the old and the new scientific mind: "hypothesis" and "synthesis," for example, "paralogy" and "analogy," "predictive" and "predicant." He uses words or phrases which have become philosophical trademarks, so to speak, defining his position in relation to Kant and Schopenhauer, for instance, through an apparent similarity which quickly turns into difference.

Why all this linguistic effort? Bachelard is, after all, presenting ideas already put forward in earlier books, but with less insistent word play. If we compare this book with his *Essai*, for example, what seems to emerge is a change in intention, the *Essai* arguing, developing a thesis, step by step and in detail, *Le Nouvel Esprit scientifique* stating and restating ideas, rephrasing and illustrating them. Indeed, it could be maintained that the whole book is summarized in one phrase: "scientific Reality is already dialectically related to scientific Reason." Perhaps this means that Bachelard has nothing new to say, or worse that the book is merely a potboiler. Dominique Lecourt, in *Bachelard ou le jour et la nuit* (1974), has been very critical of Bachelard for continually "announcing" a philosophy of science adequate to scientific thinking but failing ever to produce it. Lecourt, of course, reads with the prejudiced eye of the dialectical materialist for whom Bachelard's nonmaterialist epistemology is necessarily a "delusion." There is a grain of truth in Lecourt's criticism, in that Bachelard does not go beyond his original conception of the interdependence of reason and reality in modern science. However, it must be added that Lecourt picks on only one of Bachelard's statements of intention, that there are others. His books on epistemology may well be variations on a theme, but it is the variations that are important. When he declares that the philosopher must inflect his language, he is also stating his intention. If he makes such efforts to inflect his own language, it is not only to develop a philosophy adequate to science but to enable, even oblige his reader to gain what he will later in the book call "psychological benefit" (88). The "supple, mobile character of contemporary thought" involves, besides new concepts—"technical realism," "rational realizations"—new experiences, the "psychological dialectic" (30) of subject and object.

We shall now turn to the second of these statements chosen for discussion, "Modern science is founded upon the *project,* above the *subject* and beyond the immediate *object*. In science, whenever a subject thinks about an object, his reflection is in the form of a project." This word "project" is new in Bachelard, an inflection of language that stops the reader in his tracks: what does he mean, and where does the word come from? Bachelard has from the beginning regarded the "ambiguity" of reason and reality in modern science as having repercussions on the status of subject and object, and we have seen him try a number of ways of describing their mutual reciprocity. The object causes him fewer problems than the subject, because while one can assume that the "objects" of the new physics are generally understood to be totally unlike those of immediate experience, the very predominance of mathematics, of reason in modern science encourages the idea that the subject is sovereign, independent. This, in effect, is disproved by the discontinuity and "fundamental incompleteness" of twentieth-century science. Bachelard therefore limits the role of the subject with his conception of "approximate knowledge," insisting on the polemic of reason and reality, of consciousness and the world. However, the linguistic problem remains, since "subject" and "object" are defined as opposites, separated by an "unbridgeable gulf." A new word is needed, and "project" suggests very well both the progress, the openness of the new science, always "thrown forward," and also its polemics, reason and reality "thrown on to" each other. Besides this, it is a mathematical term, associated with projective geometry. Projections in this context are transformations of figures, the original figure and that arising from it being inseparable, since they are projections with respect to each other. Here again, "project" obviously suits Bachelard's purpose remarkably well. Yet it is surely a far cry from geometry to the relationship of subject and object. "Project" also has, in 1934, a precise philosophical sense, that given by Heidegger. Bachelard, I suggest, takes the word from Heidegger, in an intricate and revealing inflection of language.

Heidegger's work was not well known in France at this time, some ten years before Sartre's diffusion of his ideas in *L'Être et le néant* (1943), and not yet translated into French, with the exception of one article. This article, "De la Nature de la cause," the first translation of his work published in France, appears along with Bachelard's "Noumène et microphysique" in the first issue of *Recherches philosophiques* (1931–32).[1] Heidegger was therefore readily accessible to Bachelard who, given his appetite for the written word, can be assumed to have read this article. It must have struck a chord with him. "Transcendence," Heidegger writes, "defines the nature of the subject, it is the fundamental structure of subjectivity . . . *being* a subject means being an existent in transcendence and as transcendence" (93). He goes on to say that "we call *world* that *towards which* man as such operates a transcendence and we define transcendence as '*being-in-the-world*'" (95). Developing this, Heidegger explains that *"transcendence means project of the world, in such a way that the projecting being thus penetrated is as if traversed by the existent that he goes beyond"* (108), that "man founds (creates) the world only in so far as he founds himself in the midst of existence" (117). Bachelard must surely have recognized the consonance of his ideas and Heidegger's, the word "project" being especially relevant, a vivid, clearer way of expressing, on the one hand, approximate knowledge ("project of the world"), and on the other, the subject ("the projecting being . . . penetrated . . . traversed by the existent that he goes beyond"). Heidegger's "project" offers Bachelard a new way of inflecting his language, of expressing more concisely the interdependence of subject and object, the reciprocity of an active subject and a transcendent world.

The last of my chosen quotations also concerns the relationship of subject and object: "it is, moreover, when the subject thinks about an object that this subject is most likely to acquire depth." "Depth" is a puzzling word here, since Bachelard has always rejected Bergson's idea of the "deep self," insisting on the exposure of the self to an aggressive "not-self." We must be careful not to read "depth" too quickly, as a sign of recantation. On the contrary, Bachelard makes "depth" a function of the "projecting being," of transcendence, and not, as Bergson did, a retreat from transcendence. By using the word like this, he in fact defines himself *against* Bergson. Depth is not something given but something acquired, constructed, as he explains in the second of these extracts, by "objective thought . . . educated by its confrontation with organic nature." The notion of the subject acquiring depth is not so much a new idea as a new angle on an old idea, for whereas Bachelard has tended to emphasize rectification as "conquering" the object, he now attends to its other pole, to the effect of this "project" on the subject. The same shift in emphasis is seen in the argument of the last chapter of *Le Nouvel Esprit scientifique*, "Non-Cartesian Epistemology," which begins by showing how modern science invalidates Descartes's idea of objects as "simple natures" and goes on to propose a new view of the subject, a "non-Cartesian *cogito*," in keeping with the new science.

Descartes's analysis of a piece of wax in *Meditation II* embraces both subject and object, and it is because of this that Bachelard pays it such close attention. In addition, it gives him an excellent chance to show once again how "science creates philosophy." The melted wax is changed, but Descartes argues that it is still "the same wax," its "sameness" being conferred by the mind which "judges" the evidence of our senses. If he judges the wax—or anything else "exterior" to him—to exist, then, Descartes concludes, he himself must exist; if the wax is the same, he is the same, and consequently he is defined by his thought. Bachelard both accepts and rejects the Cartesian *cogito*, and this is crucial where the question of his alleged idealism and humanism is concerned. When he declares that the subject who thinks about an object acquires depth, he is going beyond Descartes, in fact, and the word "depth" is an important indicator of the non-Cartesian, Bachelardian *cogito*. He imagines how a modern scientist would examine a drop of wax, and this does indeed seem to support the primacy of reason: "the qualities of scientific reality are . . . functions of our rational methods," he declares. Where he differs from Descartes is in his conception of reason. Descartes, he suggests, values "simplicity," "unity," "constancy," and requires rather than proves that the *cogito* should have these characteristics. This "lack of impartiality" means that Descartes cannot cope with transience, with variables and diversity, which he therefore neglects, in mind as in matter. According to Bachelard, the facts of modern science prove that reason and reality are not like this. For microphysics, the drop of wax is an increasingly complex structure, not a "simple nature," and reason—the "rational methods" of mathematics, physics, and chemistry—is also complex. Rather than reason conferring "sameness" on reality, it is reality that makes reason so diverse. The passage ends with a reaffirmation of Bachelard's "epistemological postulate," the incompleteness of scientific knowledge, which, as he argued in his thesis, invalidates idealism (*ECA* 13). The scientist's rational strategies are invented in response to the problems posed by what he calls in the first of these two extracts "a reality of which he lacks full knowledge," whose "essential function" is that "it must make us think."

Reason for Bachelard is always "applied," always "transcendence" and "project." The Bachelardian *cogito* is therefore not solitary but dialectical: "I think about reality" because it is reality that makes me think. In what sense though does this *cogito* acquire depth? The thinking subject's depth is proportional to the depth of the object that is thought. But the scientific object is not deep; microphysics has proved depth to be a metaphor. Matter has no "inside," no "depth" because it is not spatial but temporal and discontinuous, because it *is* rather than *has* energy (*NES* 63–69). Pauli's exclusion principle is given particular importance by Bachelard because it puts an end to the idea that the "depth" of substance determines material properties. The electron's individuality is given by four quantum numbers which are strictly relative to the quantum numbers of other electrons; since according to the exclusion principle no two electrons can have the same set of four quantum numbers, it follows that what determines one electron's individuality is *other* electrons, numerical *difference* (*NES* 83–84). Difference is therefore a law of nature, an ontological necessity. The "scientific object" is to be defined in terms not of identity (depth) but of difference. What then of the "scientific subject," the *cogito,* acquiring depth? Depth, where the subject is concerned, has always been a metaphor, an expression of certain values—richness, diversity, possibility. However, because this "deep subject" was at the same time regarded as "pure immanence," as identity, these values could never become a fact. Scientific thinking now offers us, or so Bachelard believes, a way of turning value into fact, of escaping from, so to speak, depth identity into depth difference. Not only is the subject transcended by the object—the "not-self," the other, difference—but that object is itself defined by difference, difference which is both man-made and a law of nature. Most important of all, the subject is made different by difference, "acquiring depth" through this polemical transcendence.

Extract I
The Epistemological Break: Beyond Subject and Object in Modern Science

All educated men are, according to William James, inevitably guided by a particular metaphysics, and many have echoed his words. We would consider it more accurate to say that anyone who attempts to acquire a scientific education is dependent not just on one particular metaphysics but on two, and that these two are contradictory, even though both are equally natural and convincing, equally implicit and enduring. Let us, just for the time being, quickly put a name to these two fundamental philosophical attitudes that we find peacefully coexisting in the mind of the modern scientist; let us follow tradition and label them as rationalism and realism. Should you wish us here and now to offer some proof of this tranquil eclecticism, you have only to reflect upon the following postulate of the philosophy of science:

> Science is a product of the human mind, a product which is in conformity with the laws of thought and which is also in accordance with the outside world. Science can be considered, then, to have two aspects, the subjective and the objective, each as necessary as the other since it is equally impos-

sible for us to change anything to do with the laws of our mind and the laws of the world.[1]

What a very curious metaphysical statement this is! It can lead to a kind of redoubled rationalism which would find the laws of man's mind repeated in the laws of the world, and it can also lead to a kind of universal realism, conceiving "the laws of the mind" as part of the laws of the world and so turning them into something absolute and invariable.

The philosophy of science has not in fact cleared itself of confusion since Bouty made this statement. It would not be hard to demonstrate that, on the one hand, the most ardent rationalist allows his scientific thinking to undergo daily instruction from a reality of which he lacks full knowledge, and that, on the other hand, the most convinced realist immediately proceeds to simplify his information, precisely as though he were accepting the lessons of rationalism. This means that there is no absolute realism or rationalism in the philosophy of science, and that we must not take a general philosophical attitude as our starting point when we are considering the nature of scientific thought. Sooner or later, scientific thought will itself become the central issue of philosophical debate, and it will lead to the replacement of immediate, intuitive metaphysics by one that is discursive and objectively rectified. By following these rectifications, we will be convinced that for example a realism which has encountered scientific doubt can no longer be the same as immediate realism. We will be equally convinced that a rationalism which has corrected a priori judgments cannot remain closed, as was the case with the new developments in geometry.[2] We believe therefore that there is much to be gained from considering the philosophy of science in itself, and from examining it without any preconceptions, away from traditional philosophical terminology with its rigid requirements. Science in fact creates philosophy. The philosopher must therefore inflect his language so that it can express the supple, mobile character of contemporary thought. He must also respect that curious ambiguity which requires all scientific thinking to be translated at one and the same time into the language of realism and into that of rationalism.

---

1. Bachelard's footnote: Bouty, *La Vérité scientifique: sa poursuite* (Paris, 1908), 7. Note amended. My translation; this work has not been translated.
2. The "new developments in geometry" to which Bachelard refers are those of non-Euclidean geometry. See above, 6–7.

This metaphysical impurity ought perhaps to be regarded as a first lesson upon which we must reflect, and as a fact which we must explain, arising as it does from the twofold direction of all scientific proof, which is valid in both experimentation and in reasoning, in both its contact with reality and its reference to reason.

It is, moreover, quite easy to account for the dualistic basis of all philosophies of science; the very fact that the philosophy of science is an *applied* philosophy means that it cannot preserve the purity and the unity of speculative philosophy. Wherever the starting point of scientific activity may be, this activity will carry conviction only if it leaves its home base; *if scientific activity is experimental, then reasoning will be necessary; if it is rational, then experiment will be necessary.* To apply is always to transcend. We shall show that in the simplest scientific procedure there is clear evidence of duality, of a kind of epistemological polarization which tends to classify phenomenology under the double heading of the picturesque and the understandable, in other words, according to the double label of realism and rationalism. Turning now to the psychology of the scientific mind, we see that, if we could stand at the frontier of scientific knowledge, it would be very clear that what contemporary science is trying to achieve is the synthesis of metaphysical contradictions. Nevertheless, the direction of the epistemological *vector* seems to us to be quite plain. There is no doubt about it: it goes from the rational to the real, and not the reverse, that is, from the real to the general, as all philosophers from Aristotle to Bacon formerly maintained. In other words, it seems to us that when scientific thought is applied, we have an application that is essentially realizing. We shall try, then, to show in the course of this book what we shall call the realization of the rational, or in more general terms, the realization of the mathematical.

This need for thought to be applied is, moreover, no less active in pure mathematics, although it is rather more concealed. It introduces an element of metaphysical duality into these apparently homogeneous sciences, and with this, an excuse for realists and nominalists to go on arguing. If we are too quick to condemn mathematical realism, it is because we are seduced by the impressive extension of formal epistemology, that is to say, by the way in which mathematical concepts operate, as it were, in a vacuum. However, if we are careful not to abstract unduly from the psychology of the mathematician, we soon realize that there is more to mathematical activity than just the formal organization of schemata, and that every pure idea is doubled by a psychological application, by an example which serves as reality. Thus, as we reflect upon the work done in mathematics, we see that it always results from the extension of knowledge acquired from reality, and we see too that even in mathematics, reality is revealed in its essential function: it must make us think. Mathematical realism, in some shape, form, or function, will sooner or later come along and *give body* to thought, making it psychologically permanent, showing us the two aspects of mental activity by revealing, here as everywhere else, the dualism of the subjective and the objective.

Since our intention is to study the philosophy of the physical sciences in particular, we must try to show how the physicist realizes the rational in his experiments. This realization corresponds to technical realism and it is, we believe, one of the distinctive features of contemporary scientific thinking, so different here from the scientific thinking of the last few centuries, so far removed in particular from the agnosticism of the Positivists and from the easy tolerance of the Pragmatists, and completely unrelated, in fact, to traditional philosophical realism. What we now have is a second-order realism, realism reacting against everyday reality and contesting immediate experience, realism which is the fruit of reason that has been realized, of reason tested by experiment. The reality corresponding to it is not to be cast into the realm of the unknowable thing in itself. This reality has a quite different noumenal richness. Whereas the thing in itself is a noumenon by virtue of the fact that all phenomenal values have been excluded, it seems to us that scientific reality consists of a complex noumenal structure with the power to indicate the axes of experimentation. A scientific experiment is therefore a reason that has been confirmed. This new philosophical aspect of science is paving the way for the return of the normative to its experiments; since the need for an experiment is grasped by theory before it is discovered by observation, the physicist's job is to purify the phenomenon sufficiently for it to be possible for him to rediscover the organic noumenon. The kind of reasoning that Goblot has revealed in mathematical thinking, reasoning by construction, is now making its appearance in both mathematical and experimental physics.[3] The whole notion of the working hypothesis seems to us to be destined

---

3. Edmond Goblot (1858–1935) was a French philosopher of science, known for his books *Essai sur la classification des sciences* (Paris: F. Alcan, 1898), and *Le Vocabulaire philosophique* (Paris: A. Colin, 1901). Bachelard does not indicate the source of the idea to which he refers here.

to go into a speedy decline. Such a hypothesis can be only as real as the experiments to which it is related. It is realized. The age of unrelated, fleeting hypotheses has passed, along with the age of isolated, rather curious experiments. From now on, hypotheses are syntheses.

If immediate reality is simply an excuse to think scientifically, and no longer an object of knowledge, we must progress from description to theory, from the *how* of description to the *what* of theoretical commentary. This prolix explanation amazes the philosopher, since for him an explanation should do no more than unravel some complexity, and reveal the simple in the compound. Now, true scientific thinking is metaphysically inductive; as we shall often show in the course of this book, it reads complexity in what is simple, it pronounces laws on the evidence of facts, and rules on that of examples. We shall see how particular items of knowledge are extended and completed by modern thought and its generalizations. We shall demonstrate that there is now a kind of polemical generalization by virtue of which reason progresses from the question *why?* to the question *why not?* We shall make room for paralogy beside analogy; we shall show that in the philosophy of science, the philosophy of *why not?* has taken the place of the former philosophy of *as if.* In Nietzsche's words, anything that is decisive only comes into being *in spite of.* This is just as true in the realm of thought as it is in that of action. Every new truth comes into being in spite of the evidence, every new experiment is in spite of immediate experience.

Thus, without making any reference to that accumulation of knowledge which brings about gradual change in scientific thinking, we shall find that there is in science an almost inexhaustible source of new ideas, a kind of essential metaphysical newness. Indeed, if scientific thought can play on two opposite terms, moving between, for example, the Euclidean and the non-Euclidean, then it is, as it were, surrounded by an area of renovation. These burgeoning new languages will have very little importance if we regard them simply as modes of expression, as more or less convenient turns of phrase. If, however, we regard these expressions as more or less expressive, more or less suggestive, and as leading to more or less complete realizations, all of which we shall try to show is entirely justified, then the new, enlarged mathematics must be given a very different importance. We shall therefore insist on the dilemmatic value of these new ways of thinking, non-Euclidean geometry for instance, or non-Archimedean measurement, Einstein's non-Newtonian mechanics, Bohr's non-Maxwellian physics, or noncommutative arithmetic which could well be termed non-Pythagorean. In the philosophical conclusion to this book, we shall try to describe the non-Cartesian epistemology which seems to us to set the seal on the newness of contemporary scientific thinking.

One thing must be said in order to prevent a misunderstanding: there is nothing at all automatic about these negations, and no one should be looking for a simple means of conversion by which these new ways of thinking can be fitted, quite logically, into the framework of the old. On the contrary, what we have here is a true extension of ideas. Non-Euclidean geometry is not intended to contradict Euclidean geometry. Instead, it is a kind of adjoining factor which allows the totalization and completion of geometrical thought, and its absorption into a pangeometry. Constituted on the edge of Euclidean geometry, non-Euclidean geometry traces from without the limits of the old way of thinking, with the greatest possible precision and clarity. The same will be true for all the new forms of scientific thought that come along rather late in the day and cast a recurrent light on the obscurities of incomplete knowledge. We shall find, throughout our enquiry, the same features: extension, inference, induction, generalization, complement, synthesis, totality. All of these are substitutes for the idea of newness. This is indeed a profound newness, for it is not the newness of something discovered but rather the newness of a method.

Faced with this flowering of epistemological concepts, must we go on talking of a Reality that is distant and opaque, massive and irrational? If we do so, we are forgetting that scientific Reality is already dialectically related to scientific Reason. This dialogue between the World and the Mind has been going on for centuries and we can surely no longer talk about mute experience. If the conclusions to which a theory leads are to be denied any validity, then experience—and experiment—must provide the reasons for its opposition. The physicist is not easily discouraged by an experiment with negative results. Michelson died without having found the conditions which would, in his opinion, have rectified his experiment with regard to the detection of the ether. On the very basis of this negative experiment, other physicists decided with great insight and subtlety that this experiment, though negative in Newton's system, was positive in Einstein's. What they did, in fact, was to put into action in the realm of experiment the philosophy of *why not?* In this way, a well-made experiment will always be positive. However, this conclusion does not rehabilitate the absolute positivity of experience as such, for an experiment can be well made only if it is complete, and this is true only with regard to an experiment that follows a project which has been thoroughly worked out in accordance with a complete theory. Last, experimental conditions are the conditions of experimentation. This little distinction shows us something quite new in the philosophy of science, since it emphasizes the technical difficulties involved in realizing a preconceived theoretical project. The lessons of reality are of value only in so far as they suggest rational realizations.

We have only to think about the way science works, then, and we shall see at once that realism and rationalism endlessly offer each other advice. Neither the one nor the other, taken in isolation, can ever suffice to constitute scientific truth; in the realm of the physical sciences, there is no room for an intuition of phenomena which would establish at a stroke the foundations of reality; equally, there is no room for a rational conviction—absolute and definitive—which would impose fundamental categories on our methods of conducting experimental research. There is here a reason for methodological newness to which we must draw attention; experiment and theory are so closely related that no method, whether experimental or rational, can be guaranteed to keep its value. We can go even further and say that an excellent method will eventually cease to be fruitful if its object is not renewed.

Thus, the epistemologist must take up his position at the crossroads, and stand between realism and rationalism. It is here that he will grasp the new dynamism of these contrary philosophies, and the double movement by which science simplifies reality and complicates reason. We have, then, shortened the path that runs from explained reality to applied thought. It is here, along this short path, that we must develop the pedagogics of proof, which is, in effect, as we shall show in our last chapter, the only possible psychology of the scientific mind.

If we generalize still further, do we not see that there is something to be gained from taking the essential metaphysical problem of the reality of the external world and considering it in the realm of scientific realization? Why do we always start with the opposition between a rather vague Nature and an untutored Mind, and so confuse, without further discussion, the pedagogics of initiation and the psychology of culture? How dare we step outside ourselves and recreate the World in one brief hour? How can we assert that a simple, naked self can be grasped independently of its essential action in objective knowledge? We shall cease to be so absorbed by these elementary questions if we simply consider, in addition to the problems of science, the problems posed by the psychology of the scientific mind, and regard objectivity as a difficult pedagogical task and not as something immediate and given.

Furthermore, it is perhaps in scientific activity that the two aspects of the ideal of objectivity are most clear, that is, the real and social value of objectivation. To quote Lalande, the aim of science is not simply "the assimilation of one thing and another, but first and foremost the assimilation of one mind and another." Without this assimilation, there would, in a manner of speaking, be no problem. Were we entirely autonomous beings who confront a highly complex reality, our quest for knowledge of that reality would lead us to those aspects that we find picturesque or evocative: *the world would be our representation*. If, on the other hand, we were entirely dependent upon society, we should seek knowledge in the general, the useful, the accepted: *the world would be our convention*. In actual fact, scientific truth is predictive, or rather, it is predicant. We call upon minds to converge by preaching the good news of science, by passing on at one and the same time both a thought and an experience, linking thought to experience in verification: *the scientific world is, therefore, our verification*.[4] Modern science is founded upon the *project*, above the *subject* and beyond the immediate *object*. In scientific thought, whenever a subject thinks about an object, his reflection is in the form of a project.

We should, moreover, be much mistaken if we were to try to argue from the fact that real discoveries are very rare in these Promethean labors; in the very humblest kind of scientific thinking, this indispensable theoretical preparation is perfectly clear. In a previous book, we declared without any hesitation that reality has to be demonstrated, for it cannot be displayed. This is particularly true when we are working on organic phenomena.

---

4. A note is required about the translation of the last few lines. Bachelard plays with words here, moving from the readily accessible idea that scientific truth is "une prédiction" to the more striking declaration that it is "une prédication." "Prédication" in French means a sermon, or the act of preaching. My translation attempts to echo Bachelard's wordplay by turning his nouns into adjectives in English. He goes on to exploit the notion of preaching in the phrase "nous appelons les esprits à la convergence en annonçant la nouvelle scientifique." Literally translated, this would be "we call upon minds to converge by announcing the scientific news." This however loses the force of Bachelard's association of "preaching" with "la nouvelle," which refers not just to news but to "la bonne nouvelle," the good news, the gospel.

Indeed, when an object is presented as a complex set of relations, we must use a multiplicity of methods in order to apprehend it. Objectivity cannot be separated from the social aspects of proof. Objectivity can be attained only through the detailed and discursive exposition of a method of objectivation.

We believe, then, that fundamental to all objective knowledge there must be this preliminary demonstration, and this belief is surely unassailable where science is concerned. Observation already requires a *set* of safety measures which will make us think before we look, or which will at least correct our first impression of what we see, with the result that our first observation is never the right observation. In science, observation is always polemical; it either validates or invalidates a previous thesis, a preliminary schema, a level of observation; it displays a fact by demonstrating it; it orders appearances into a hierarchy; it transcends the immediate; it reconstructs reality after first reconstructing its schemata. When we go from observation to experimentation, the polemical character of knowledge will naturally be even clearer. Phenomena must now be carefully selected, filtered, and purified; they must be cast in the mold of scientific instruments and produced at the level of these instruments. Now, instruments are just materialized theories. The phenomena that come out of them bear on all sides the mark of theory.

Thus the relation between the phenomena and the noumena of science is no longer to be seen as some remote and rather indolent dialectics; it is, instead, an alternating movement which always tends toward the effective realization of the noumenon, after first rectifying a few projects. In this way, then, the true phenomenology of science is in fact essentially a phenomeno-technique. It reinforces what can be glimpsed just beyond appearances. It is instructed by what it has constructed. Thaumaturgic reason traces its own framework and limits in accordance with the schema of the miracles it performs. Science calls a world into being, not through some magic force, immanent in reality, but rather through a rational force, immanent in the mind. Whereas reason was, in the early days of science, formed in the image of the world, now, in modern science, the aim of mental activity is to construct a world in the image of reason. Scientific activity realizes, in the full meaning of the term, rational groups.

Le Nouvel Esprit scientifique (1934; Paris: Presses Universitaires de France, 1973), 5–17.

Extract II
Toward a Non-Cartesian Epistemology

Once we have understood that modern mathematical thinking has gone far beyond the science of spatial measurement it originally was, and that the science of relations has now made very great advances, we shall realize that mathematical physics provides us day by day with ever more numerous axes for scientific objectivation. We shall find that the stylized nature of the laboratory prepared by mathematical schemata is much less opaque than the nature which appears to immediate observation. And vice versa, when objective thought is educated by its confrontation with organic nature, it will immediately reveal remarkable depths within itself, by virtue of the fact that this kind of thinking is perfectible, rectifiable, and able to suggest complements of itself. It is, moreover, when the subject thinks about an object that this subject is most likely to acquire depth. Instead of following the metaphysician into his "stove-heated room," we may well prefer to follow the mathematician into his laboratory.[1] Indeed, we shall soon see Plato's warning words pinned on the door of the physics or the chemistry laboratory: "nobody untrained in Geometry may enter my house."

Let us compare, for example, Descartes's observation of a piece of wax with the way contemporary microphysics would deal with it: let us see the different results obtained for both an objective and a subjective metaphysics of substance.[2]

For Descartes, the piece of wax symbolizes clearly and simply the transience of material properties. Nothing is permanent here, be it a property of the whole or an immediate sensation. We have only to place the wax near a fire and we shall see its consistency, form, color, unctuousness, and smell all waver and be transformed. In Descartes's eyes, this indefinite

---

1. This phrase "stove-heated room" is a reference to the *poêle* which Descartes describes at the beginning of Part II of his *Discourse on Method*. Shut up alone all day in this hot room, in the winter of 1619–20 at the beginning of the Thirty Years' War in Germany, he spends the time thinking and eventually formulates the principles of his method.

2. Bachelard is referring to Descartes's *Meditation II*. See the translation by Elizabeth S. Haldane and G. R. T. Ross, *The Philosophical Works of Descartes*, 2 vols. (Cambridge: Cambridge University Press, 1911; corrected reprints, 1931–34); rpt. New York, 1955, I, 154–57.

experiment proves the indefiniteness of objective qualities. It teaches us to doubt. It leads the mind to foresake the experimental knowledge of bodies, for it is harder to know these than it is to know the soul. Were our understanding unable to find within itself this knowledge of extension, then the substance of our piece of wax would vanish along with imagination's airy dreams. The piece of wax is maintained simply and solely by *intelligible* extension, since it is very likely to increase or decrease in size according to circumstances. This refusal to regard experience as the basis of thought is in fact complete and final, despite the return to the study of extension. Descartes rejects, from the very beginning, any kind of progressive experience, any means of classifying the various aspects of diversity and of revealing its riches, and any way of immobilizing the variables of a particular phenomenon in order to set them apart from one another. His ambition was to find the object's simplicity, its unity and its constancy, and find them straightaway. The very first failure to do so meant that *everything* was brought into doubt. It was not understood that a factitious experience can play a very important part in coordinating, nor that thought linked to experience can restore the phenomenon's organic character, making it therefore something whole and complete. On the other hand, the refusal to submit without resistance to the lessons of experience meant that it was quite impossible to see that the mobile character of objective observation found its immediate reflection in the parallel mobility of subjective experience. If the wax is changing, then I am changing; I change along with my sensation, for at the moment in which I think this sensation, it constitutes my entire thinking, for feeling is thinking in the widest Cartesian sense of the *cogito*. Yet Descartes secretly believes in the reality of the soul as substance. He is dazzled by the instantaneous light of the *cogito* and so he does not cast doubt on the permanence of the *I* that is the subject of *I think*. Why is it the same being that experiences hard and soft wax when it is not the same wax that is experienced on two separate occasions? If we were to express the *cogito* in the passive as *cogitatur ergo est*, would the active subject then evaporate along with his fleeting and indistinct impressions?

Descartes's lack of impartiality where subjective experience is concerned will perhaps be clearer when we come to live out an objective scientific experience with greater commitment, when we agree to live according to the rhythms of thought, in the rigorous equation of thought and experience, noumenon and phenomenon, far from the deceptive lures of objective and subjective substance.

The New Scientific Mind

Let us now consider contemporary science as it strives toward progressive objectivation. The physicist does not take wax that is fresh from the hive but rather wax that is as pure as possible, that is chemically well defined and isolated as the result of a lengthy sequence of methodical manipulations. The wax chosen is, so to speak, a precise *moment* in the method of objectivation. It bears no trace of the perfume of the flowers from which it was gathered; instead, it is the proof of the care with which it has been purified. It has been realized, we might say, by factitious experience. Without this factitious experience, wax of this kind, in this pure form which is not its natural form, would never have come into existence.

The physicist first melts a tiny fragment of this wax in a little dish and then solidifies it, slowly and methodically. Fusion and solidification are in fact obtained very gradually by means of a minute electric furnace whose temperature can be controlled with the greatest possible precision by varying the intensity of the current. The physicist becomes, therefore, *the master of time,* for the effective action of time depends on thermal variation. He obtains in this way a drop of wax that is very regular, not just in its form but in its surface structure. The book of the microcosm has now been written, and it remains for us to read it.

In order to study the surface of the wax, a beam of monochromatic X rays is directed at it, in accordance here again with a very precise technique and, of course, without the use of natural white light, which prescientific times postulated as simple in nature. Thanks to the slowness of the cooling process, the molecules on the surface of the wax are arranged in relation to the general surface. This arrangement will, with respect to the X rays, determine diffractions which produce spectrograms similar to those obtained by Debye and Bragg with regard to crystals.[3] We know that these spectrograms, which Von Laue predicted, have completely renewed crystallography, in that they allowed the internal structure of crystals to be inferred.[4] In exactly the same way, the study of our drop of wax will en-

---

3. Peter Joseph William Debye (1884–1966), the Dutch-born physicist, worked in particular on the interaction of radiation with matter, carrying out theoretical investigations of X-ray scattering. William Henry Bragg (1862–1942), the British physicist, also worked on X rays and in the field of crystal structure analysis, discovering X-ray diffraction by crystals in 1912.

4. Max von Laue (1879-1960), the German theoretical physicist, established the new field of X-ray structural analysis, working on its fundamental theory. He contributed towards the discovery of X-ray diffraction crystals in 1912 by William Bragg and his son Laurence.

tirely renew our knowledge of the surfaces of matter. What an abundance of thoughts this prodigious epigraphy of matter must surely bring us! In Jean Trillat's words: "A very large number of surface properties, such as capillarity, unctuousness, adhesion, adsorption, and catalysis, are all dependent on the phenomena of the arrangement of molecules with respect to each other."[5] It is here, in this thin film, that the relations with everything external bring into being a new physical chemistry. It is here that the metaphysician can best understand how relations determine structure. If we take diagrams of molecular patterns and go deeper and deeper into our drop of wax, the arrangement of the molecules gradually disappears, the microcrystals lose their sensitivity to surface action, and we reach a state of total statistical disorder. In the privileged zone of molecular arrangement, we have, on the contrary, very clearly defined phenomena. These are due to discontinuities in molecular fields on the surface separating two different mediums, in the area of material dialectics. In this intermediate region, some rather odd experiments are possible, which bridge the gap between physical and chemical phenomena, allowing the physicist to act upon the *chemical nature* of substances. Thus Trillat draws attention to experiments on the stretching of colloidal gels. By means of purely mechanical traction, notable differences can be produced in X-ray diagrams. Trillat draws the following conclusion: "This is related to mechanical properties and also to the adsorption of dye, according to whether or not matter is given direction by traction: this may well offer a new and unlooked-for way of working on chemical activity."[6]

If we use mechanical means to work on chemical activity, we are, in some ways, following a Cartesian ideal. However, this action is so obviously constructive and factitious, and its orientation toward complexity is so plain, that we cannot but see this as new proof of the scientific extension of experience and a new opportunity for a non-Cartesian dialectics.

Can we be sure, moreover, that crystallization can take place in the absence of directional forces? If we imagine that this crystallization is pro-

---

Work on "Laue's diagrams" helped William Bragg develop his work on crystal structure analysis.

5. Bachelard's footnote: Jean Trillat, "Étude au moyen des rayons X des phénomènes d'orientation moléculaire dans les composés organiques" ["A Study of Molecular Arrangement in Organic Compounds by X-ray Methods"], in *Activation et structure des molécules* (Paris, 1928), 461. My translations; this book has not been translated.

6. Ibid., 456.

The New Scientific Mind

duced by forces which are essentially internal and substantial, and neglect the directive action that comes from the outside, then we are obeying the dictates of realism. It is indeed most remarkable to see that surface crystallization depends first and foremost on discontinuities, so much so that we may speak of substances which are crystallized superficially in a direction perpendicular to the surface, while remaining amorphous in a direction parallel to the surface. We thus obtain crystal lattices with clearly specified implantations. This new kind of crystalline "culture" has already taught us much about molecular structure.[7]

If we now reflect upon the number of techniques, hypotheses, and mathematical constructions that combine in these experiments on our drop of wax, we cannot fail to see that a metaphysical critique of the Cartesian kind is entirely ineffective. What is transient is just unconnected circumstances, not those coordinated relations which express material qualities. We have only to disentangle circumstances that are *naturally* entangled and reality will be truly organized. The qualities of scientific reality are therefore first and foremost functions of our rational methods. If we are to constitute a clearly defined scientific fact, we have to make use of a coherent technique. Scientific action is essentially complex. The active empiricism of science develops through complex, factitious truths, and not through those that are clear and adventitious. Innate truths have, of course, no place in science. Reason has to be formed in exactly the same way that experience has to be formed.

Thus, the objective meditations that take place in the laboratory will lead us toward a progressive objectivation in which, at one and the same time, a new experience and a new kind of thought are realized. This kind of meditation is very different from subjective meditation with its thirst for knowledge which is clear and complete, different, too, because of its very progress and because of its need for a complement whose existence it always supposes. The scientist leaves his laboratory in the evening with a program of work in mind, and he ends the working day with this expression of faith, which is daily repeated: "Tomorrow, I shall know."

Le Nouvel Esprit scientifique (1934; Presses Universitaires de France, 1973), 170–77.

---

7. Bachelard's footnote: cf. Jean Thibaud, "Études aux rayons X du polymorphisme des acides gras" ["Studies of the Polymorphism of Fatty Acids by X-ray Methods"], in *Activation et structure des molécules*, 410 ff.

The *schola quantorum,* as Bachelard calls it, is not an elitist establishment for the mathematical few; it is for all of us, though we can perhaps never really understand the new science. What matters is that we should struggle to understand, that we should make the epistemological break even if its implications remain beyond our mental grasp. But why does it matter, why not rest content with what we have and what we know? Why not leave science to the scientists, and get on with our own lives which, it is perfectly true, are untouched by Pauli, Heisenberg, and the like? If Bachelard insists on teaching us the new science, it is not simply because it is there, because these are facts of which we ought, as intelligent creatures, to be aware. The new science teaches us about ourselves, and this is an increasingly important theme in his work. He has a number of aims in mind when he writes, which are reflected in his different statements of intention. One of the most revealing is made in *Les Intuitions atomistiques*, published in 1933, a year before *Le Nouvel Esprit scientifique*, when he defines his aim in his epistemological work as the "task of catharsis" (14). Why this word "catharsis"? Aristotelian connections can be ruled out as there is obviously no reference to emotion. But "catharsis" was also used to describe psychoanalysis in its early days. Freud, for instance, called his psychotherapy "Breuer's cathartic method," continuing even in 1909 to refer to it as his "cathartic procedure."[2] "Catharsis" here means healing, restoring to health. When Bachelard describes his aim as the "task of catharsis," the implication is that he is seeking to cure his readers. But of what? Of the pressures of immediate reality and the privilege of "good sense." Life, with its habits, its enforced and prized identity, is pathogenic, and this idea recurs in *Le Nouvel Esprit scientifique*: ordinary life and thought lead to mental "ankylosis" (43), to "psychological hardening" (87). Even more strikingly, he uses the phrase the *"geometric unconscious"* (41), and describes modern science as the "psychoanalysis of Euclidean impulses" (42). *Homo faber* is not just wrong, he is diseased. The healthy mind is the "new scientific mind." The "supple, mobile character of contemporary thought" is reformulated toward the end of the book as the "mobility of healthy methods" (140), and the "project" of modern science is held to offer all of us "excellent mental hygiene" (88). Bachelard writes as a philosopher, a teacher, and also as a healer. However odd it may seem, this therapeutic intention is an important element in his work: *homo aleator* is a fact of modern science, but he is also for Bachelard a value, the model of the healthy, creative human being we should all become.

Chapter 4

Time, Consciousness, and Discontinuity: 1936

> This century has seen a psychic revolution; human reason has weighed anchor, the mind's voyage has begun, and knowledge has left the shores of immediate reality. Is it not therefore anachronistic to cultivate a taste for a safe harbor, for certainties and systems?
> — *L'Engagement rationaliste*

With *La Dialectique de la durée*, published in 1936, Bachelard turns from the exploits of "thaumaturgic reason" in twentieth-century science to what he describes as the "adventures of consciousness," exhorting us in the first of these two extracts "(Cogito)^3," from chapter 6, "Temporal Superimposition," to "live temporally at the power of three, at the level of *cogito* cubed." Let us live dangerously, on the frontiers of thought; but why? Why this odd notion of (cogito)^3, *I think that I think that I think*? Not only is it very difficult to maintain this level of consciousness, breaking as it does with everyday life and with scientific thinking, but it may well seem dangerous, since we lose our familiar sense of ourselves as material, historical, and rational creatures. Yet for Bachelard, this danger is creative. (Cogito)^3 shows us our potential, what we ought to be, and paradoxically, it brings great happiness and "repose."

The notion of repose is at the heart of this book. The very first sentence presents its aim as "a propaedeutics for a philosophy of repose." This surely comes as something of a shock, after all Bachelard's lessons in polemical thought. However, as both these extracts show, Bachelard does not contradict his established ideas and values but considers them from another point of view. (Cogito)^1 corresponds, in effect, to immediate, everyday reality, to a closed, Bergsonian world of ends and means, where we are created by the objects we use, and Bachelard gives it the

Aristotelian label of "efficient cause." (Cogito)² —Aristotle's final cause— corresponds to second-order reality, polemical thought against a transcendent "not-self" that defines us. (Cogito)³, as the phrase "formal cause" suggests, is the pure consciousness that we alone form, that is our form, our being. In interpreting this idea, it is important to remember that (cogito)³ is grounded on (cogito)², that it is not unpolemical but nonpolemical— to use a favorite Bachelardian formula—that is to say, beyond polemics. In (cogito)³, we break with polemics, with rectification, with what he has called "the rhythm of mathematical approximation" (*ECA* 299), and we establish ourselves in pure thought, in that pure consciousness which is consciousness not just of progress in our own being but of the frontiers of our being.

Bachelard describes this pure consciousness as follows. He refers in the first of these two extracts to "this strictly tautological psychology in which being is really and truly self-concerned," and a few lines later to "this becoming . . . peripheral to the becoming of things . . . independent of material becoming . . . this formal becoming rises above and overhangs the present instant." It might appear that these statements contradict each other. Indeed, the word "tautology" does suggest identity, and from "self-concerned" we infer self-sufficiency, so how can Bachelard speak of "becoming"? Again, it must be remembered that this "self-concern" is not immediate, that it comes after polemical thought, that tautology has to be conquered, endlessly reconstructed against the claims of the "other," the transcendent "not-self." This explains Bachelard's use of the term "instantaneity." (Cogito)³ implies not identity, not continuity, but discontinuity, difference. Here, we are conscious of ourselves as project, as pure project, transcended not by the world, the not-self, but by our own self, by difference latent within us, by the other that is our self.

Bachelard's description of this pure project is once again highly idiosyncratic—this "formal becoming," he writes, "can shoot up like a rocket . . . high above ordinary psychic life"—but it is also apt, a vivid description of a severed, momentary consciousness. As in *Le Nouvel Esprit scientifique*, he associates this idea of project with that of depth, formulating the concept of "vertical time." And as in *L'Intuition de l'instant*, time is discontinuous, but we note that this vertical time is constituted not as it was previously by instants of "nascent knowledge" but by instants of severance. Beyond the instant in which "only ideas are born" (*II* 19). Bachelard now postulates an instant where "we watch the person being born." In vertical time, we are truly free, truly ourselves; in the "dialectics of duration," we pursue a dialogue with ourselves, with a self that is ever renewed, ever different. This special kind of self-reference, which is dialectical, rhythmic, coherent, and cohesive, will, he believes, bring great repose.

How does Bachelard come to formulate this notion of "pure consciousness"? It is compatible with his previous work, this we can see, but it is nonetheless apparently out of line with what we would consider his real preoccupations. The opening paragraphs of this first extract here provide a clue: in recounting his dream—with such delightfully inventive humor—he is in fact putting Bergson to the test, for Bergson regarded dreams, where we are cut off from external things, as giving us privileged access to the "deep self," to pure consciousness (1976: 85–109 [1920: 84–108]). Bachelard's own dream does not reveal duration or continuity, but instead complex layers of time, and what he calls "temporal superimposition." The continuity of waking life, of transitive time, is disrupted when we sleep; verbal and visual time are "disengaged," shown to be independent and discontinuous. Bergson was wrong: pure consciousness is consciousness of discontinuity. There is surely no evidence for those who, like Jacques Gagey, see in Bachelard "an implicit Bergsonism" (1969: 62). He has been from the beginning and he will remain Bergson's "tenacious adversary," to quote Jean-Claude Margolin (1974: 27), *La Dialectique de la durée*, as the title shows, being a further stage in his polemic with Bergson. The first chapter in the book is devoted to a detailed refutation of Bergson's discussion of *le néant*, "nothingness." Here, Bachelard tackles him from a rather different point of view, and what he says is striking proof of the coherence of his own thinking: his idea of pure consciousness, "full of lacunae, broken up by great intervals of time"—consciousness of severance—rests on a conception of "nothingness" that stems from his epistemology.

Bergson argues at length in the fourth chapter of *L'Évolution créatrice* that there can be no *néant*, but only being, only fullness and duration, since in trying to imagine nothing, not only do we remain conscious of ourselves but we have to imagine something in order to annihilate it. Nothingness is therefore a "pseudo-idea"; it is an illusion of our own making, derived from everyday life where our actions do indeed appear to create something from nothing, something we want but do not have. For Bergson, nothing is new, nothing is created, reality is fullness. All Bachelard's work on science forbids such a view. As he declares in the foreword to *La Dialectique de la durée*, scientific thinking proves the "negating powers of the mind" (vi): it is polemical, destroying before it creates, annihilating appearances, negating first-order reality, its value being in proportion to the errors it has overcome (14–16). To think is to negate, and since both consciousness and time are for Bachelard a function of thought, nothingness is fundamental to our being, not a flaw but a fact, an ontological necessity (29). It is the guarantee of progress, of becoming. It is clear, therefore, that Bachelard's wish to explore what he calls "the psychology of annihilation" in this book is entirely in line with his previous preoccupations.

Bachelard is always alert to other people's ideas, through which he refracts his own pattern of thought, and so it is in order to pursue his study of this "psychology of annihilation" that he discusses in chapter 8 of *La Dialectique de la durée* the "rhythmanalysis" of the Brazilian philosopher Pinheiro dos Santos. The second of these two extracts, "On Poetry and the Dialectics of Duration," is taken from this chapter, and chosen because it is here that for the first time Bachelard describes how he reads poetry. He finds "resonances" in dos Santos because, as he explains, "rhythmanalysis" is a form of psychotherapy, aiming to cure disturbed people by renewing their awareness of natural or biological rhythms. It differs from and goes beyond psychoanalysis by trying to establish not just a balance between consciousness and the unconscious but a "double movement," a rhythmic interchange between the two poles of the psyche. Furthermore, "rhythmanalysis" is a theory based on modern physics, where matter is energy, temporal and rhythmic. Everything is rhythmic, says Bachelard, we walk on vibrations, sit on vibrations, live in vibrations, and are ourselves vibrations (131). Dos Santos' attraction for him is easily guessed at, an opportune reminder of his own therapeutic intentions, but he also disagrees with him, in particular on the subject of the unconscious which, Bachelard believes, ties us to the past, imprisons us, and destroys our progress. He therefore goes beyond dos Santos' dialectics of consciousness and the unconscious to the dialectics of pure consciousness and the "dialectics of duration," to thought time and the rhythm of being and nothingness. It is admittedly very difficult to maintain this level of consciousness, this (cogito)[3] so necessary to our well-being. Here, Bachelard suggests that poetry can help us achieve this restful, vibrant self-reference. He is concerned not at all with the meaning of poetry, but only with the effect of poetry on the reader. Poetry is presented as a structure of ambiguity, and reading a poem as experiencing this ambiguity, superimposing images and interpretations. But this superimposition implies discontinuity, consciousness of different

Time, Consciousness, and Discontinuity

meanings: when we read, we refuse one meaning for another, we resist one image and prefer another, only to destroy it, too, in its turn. Bachelard points out that when we read a poem, we are in fact *dissociating* interpretations, thus exploring and maintaining difference. It is for this reason that he proposes poetry as a "model of rhythmic life and thought." A poem's complex structure gives the mind the close pattern of reference points it needs, and without which there is indeed no thought and no consciousness. Moreover, because it is the reader who chooses between the different meanings of an image, poetry also offers him the experience of creative self-reference. When we read a poem in this way, we accede to pure consciousness, to that active, vibrant repose which is pure project. The poet, like the mathematician, frees us from the prison of the conventional and the identical, so releasing us into difference, into an openness of being.

Extract I

(Cogito)[3]

Let us now offer an example of superimposed time taken from our own experience; it comes from a dream in whose structure we can distinguish between the different kinds of superimposed time and the parts they each play. I had bought a house, and I fell asleep one night thinking of some of the things still to be done. In my dream, my continuing worries meant that I met the owner of my old home. I took the chance, therefore, of telling him about my new acquisition. I spoke kindly as I was about to give him a piece of bad news; could anyone fail to regret the loss of a philosophical tenant, one who is ever content and uncomplaining, who combines all the integrity of a moral principle with the hermit's frugality! Then slowly, and with a skill that revealed the striking continuity of capitalist time within me, of which I was entirely unaware, I suggested to my landlord all the many ways in which we might mutually agree to end the contract binding us. I spoke at some length, with sweet words of courtesy and persuasion. My little speech was well organized: the fact that my aim was clear meant that my arguments were produced at exactly the right moment. Suddenly, I looked at the person to whom I was speaking; he was now listening to me very calmly, and he was not my landlord. He had certainly been my landlord to begin with, this I realized through some strange kind of recur-

rence; he had then been my landlord in his younger days, and afterward had turned into someone progressively more different until I suddenly realized that I was telling all this to a complete stranger. This piece of bungling on my part annoyed me so much that I flew into a temper over this fresh evidence of my absentmindedness and of the temporal discords that I had allowed to occur within me as a result of my having "superimposed time." I was awoken by the anger that so often, in our dreams, disrupts and shatters time.

Do we need any further proof that verbal time and visual time are, in fact, only superimposed, and that, in all our dreams, they are independent of each other? Visual time passes more swiftly, and it is for this reason that they fall out of step. Had I been able not only to free myself of all my financial worries but also to speed up what I was saying, I should have maintained complete synchronism with what was happening visually; dreams are indeed extremely changeable horizontally, that is to say, along the plane of the normal, everyday incidents of life, yet even so, my dream would at least have retained its vertical coherence, that is to say, the form of normal, everyday coincidence. In my conversation with the stranger who took my landlord's place, I should have chosen words which were *appropriate*. I should not have *continued* with my story: I should have *modified* my confidence the moment my confidant began to change.

If we agree to analyze complex dreams from the standpoint of these differences in temporal behavior, we shall see that there is much to be gained from the concept of superimposed time. Many dreams will seem incoherent simply because there is a sudden, instantaneous loss of coordination between the different times that we are experiencing. It would appear that, when we sleep, our different nerve centers pursue their own, autonomous development and that they are in effect time detectors, each with its independent rhythm. Let us say in passing that these isolated detectors are particularly sensitive to temporal parasites. Indeed, often in my peaceful slumbers I have the feeling that parts of my brain are crackling, as if the cells were exploding or some kind of partial death were rehearsing its disasters. If we consider time in the context of cellular activity, we must see that it is very close to the time of a moth or an amoeba, and any coincidence there is exceptional. When, like a beehive, the whole of our brain comes to life, it is statistical time that will restore both regularity and slowness. Moreover, in waking life, reality offers grounds for agreement. Reality makes what we see wait for what we say, and as a result of this we have objectively coherent thought, a simple superimposition of two terms which mutually confirm one another and usually suffice to give an impression of objectivity. We say what we see; we think what we say: time is truly vertical and yet it flows, too, along its horizontal course, bearing with it the different forms of our psychic duration, all according to the same rhythm. Dreaming is the very reverse of this, for it disengages these different kinds of superimposed time.

We have probably now adduced sufficient evidence, evidence drawn moreover from sufficiently heterogeneous sources, to have some kind of certainty that with this temporal superimposition, we are touching upon a natural problem. Let us suggest, then, how we personally would wish to direct research in order to solve this particular problem.

The temporal axis that lies perpendicular to transitive time, to the time of the world and of matter, is an axis along which the self can develop a formal activity. It can be explored if we free ourselves from the matter which makes up the self, and from the self's historical experience, in order to consolidate and sustain aspects of the self which are progressively more formal, and which are indeed the truly philosophical experiences of that self. The most general and the most metaphysical method of approach would be to build up tiers of different kinds of *cogito*. We shall return later to particular examples of this that are closer to everyday psychology. Let us now turn without further delay to our own concern with creating a compound metaphysics, a compound idealism, which will put in the place of *I think, therefore I am* the affirmation that *I think that I think, therefore I am*. We can see even now that existence as it is averred by the *cogito cogitem* will be much more formal than existence as it is implied by thought alone; if eventually we can manage to reveal what we really are when we first establish ourselves in the *I think that I think*, we shall be less inclined to say that we are "a thing which doubts, understands, conceives, affirms, denies, wills, refuses, which also imagines and feels."[1] We shall thus avoid settling into a phenomenal existence which needs permanence in order to be confirmed. The Cartesian *cogito* is necessarily discursive, for it is entirely horizontal, and this fact has been made abundantly clear by Teissier du Cros in an article which we consider to be quite unusually profound. He argues that

---

1. Haldane and Ross, *The Philosophical Works of Descartes* (New York, 1955), I, 153. Bachelard is again referring to Descartes's *Meditation II*.

between the *I* and the *am,* there is the relation of affirmation and confirmation. Where the self is concerned, the judgment of existence is, in the end, a matter of *repetition:* if we take the specific experience of the self in the context of ordinary, everyday life, and consider it in relation to the specific experience of things, within the same context, we shall find it to be identical with this.²

If, however, we can rise to the I think that I think, we shall already be free of phenomenological description. If, continuing a little further, we reach the I think that I think that I think, which will be denoted by (cogito)³, then separate, consecutive existences will appear in all their formalizing power. We have now embarked upon a noumenological description which, with a little practice, will be shown to be exactly summable in the present instant, and which, by virtue of these formal coincidences, offers us the very first adumbration of vertical time.

What we are doing here is not in fact thinking ourselves thinking something, but rather thinking ourselves as someone who is thinking. Indeed, with this formalizing activity, we watch the person being born. This formal personalization takes place along an axis whose direction is entirely opposite to substantial personality, the personality that is supposedly deep and original, although in actual fact it is encumbered and weighed down by passion and instinct, and, moreover, imprisoned by transitive time. Along the vertical axis we are suggesting, being is spiritualized in accordance with the degree of its awareness of this formal activity, of the power of the cogito it is using, and also of the exponent of the compound cogito which is as far as it can go in its attempted liberation. Were we to overcome the difficulties surrounding the first severance, and then reach, for example, (cogito)³ or (cogito)⁴, we should immediately recognize the great value of this strictly tautological psychology in which being is really and truly self-concerned, that is to say, the value of repose. Here, thought would rest upon itself alone. I think the I think would become the I think the I, this being synonymous with I am the I. This tautology is a guarantee of instantaneity.

How is it that this sequence of forms can acquire a specific temporal character? It can do so because it is a becoming. This becoming is, of course, peripheral to the becoming of things, and it is independent of

2. Bachelard's footnote: Ch. Teissier du Cros, "La Répétition, rythme de l'âme, et la foi chrétienne" in *Études théologiques et religieuses* (Montpellier, May 1935). My translation.

material becoming. Clearly, then, this formal becoming rises above and overhangs the present instant; it is latent in every instant that we live; it can shoot up like a rocket, high above the world and nature, high above ordinary psychic life. This latency is a rigorously ordered sequence. Any alteration in the order of the various tiers is inconceivable. It is, we can be sure, a *dimension* of the mind.

Someone is bound to ask whether this dimension is infinite. To draw such a conclusion would be to yield far too quickly to the seductions of logic and grammar. We therefore refuse to go on lining up our subjunctives indefinitely. In particular, we refuse to imitate those writers who talk so very vaguely about *knowledge of knowledge* . . . precisely because the subjective factor of formalization is not always clearly implicit in *knowledge of knowledge* . . . , in $(knowledge)^2$. We ourselves have found it exceedingly difficult, psychologically speaking, to attain to $(cogito)^4$. We believe that the true region of formal repose in which we would gladly remain is that of $(cogito)^3$. From our researches into compound psychology, which we shall be outlining later on, we see that the power of three corresponds to a state which is sufficiently new for us to have to exert ourselves considerably, and for a long time, before we can go beyond it and proceed with our construction. $(Cogito)^3$ is the first really unballasted state in which consciousness of formal life brings us a special kind of happiness.

These different temporal levels can be rather schematically and crudely characterized by a number of mental or spiritual causalities. Thus, we consider that if $(cogito)^1$ is implied by efficient causes, then $(cogito)^2$ can be ascribed to final causes, since if we act with an end in view, we are acting with a thought in view, while being at the same time conscious that we are thinking that thought. Only with $(cogito)^3$ will we find formal causality in all its purity. This division into things, aims, and forms will of course seem artificial to any linear psychology that seeks to place all entities on the same level, making them part of a single reality, beyond which there can only be dreams and illusions. Yet the discursive, hierarchical idealism that we are proposing is not limited to this one realist view of things. If we take Schopenhauer's fundamental axiom as our starting point and say that the world is my representation, then it is acceptable to attribute ends to the *representation of representation,* while forms that are constituted in those mental activities which imply both things and ends must be attributed to the *representation of the representation of representation.* Psychologically speaking, if we follow the axis of liberation and free ourselves of all things material, we shall then no longer determine our own being by referring to things or even to thoughts, but rather by reference to the form of a thought. Mental and spiritual life will become pure aesthetics.

Finally, the time of the person, vertical time, is discontinuous. Were we to attempt to describe the passage from one power of the *cogito* to the next as a continuous process, we should realize that we were placing it along the familiar axis of time, and by this we mean vulgar time. This would lead to a complete misunderstanding of temporal superimposition. We should be starting from the mistaken belief that all psychological description is historical, and that it is because we are following the hands of the clock that we can successively *think*, then *think that we think*, and then *think that we think that we think*. We should be disregarding the principle of the fundamental instantaneity of all well-ordered formalizations. If we take psychological coincidences not simply in the instant but also in their hierarchical form, we shall receive far more from them than potential linear development. We remain entirely convinced that the mind thrusts far beyond the line of life.

Let us live temporally at the power of three, at the level of *cogito* cubed. If this third state is examined temporally in relation to the first state, in relation, that is, to transitive time, it will be full of lacunae, broken up by great intervals of time. Here, quite unmistakably, time is dialectical. Any continuity must, once again, be sought elsewhere, in life, perhaps, or perhaps in primary thought. Yet life and primary thought are so devoid of interest for anyone grown familiar with that formal state in which we seek rest from life and thought, that as a result, this purely material continuity will pass unnoticed. What we need, then, is some kind of rational coherence to replace material cohesion. In other words, if we would like there to be a purely aesthetic kind of thought, we must transcend the dialectics of time by means of forms, by means of the attraction of one form to another. Were we to retain our ties with ordinary life and thought, this purely aesthetic activity would be entirely fortuitous, lacking any coherence or any duration. In order that we may have duration at the third power of the *cogito*, we must seek reasons for restoring the forms we have glimpsed. This will come about only if we can teach ourselves to formalize a wide variety of psychological attitudes.

La Dialectique de la durée (1936; Paris: Presses Universitaires de France, 1950), 96–103.

Extract II

On Poetry and the Dialectics of Duration

*R*hythmanalysis *will look anywhere and everywhere in order to discover new opportunities for creating rhythm. It firmly believes that there is a definite correspondence between natural rhythms, or alternatively, that they can easily be superimposed, one rhythm imparting momentum to another. Rhythmanalysis forewarns us, then, of the dangers of living at odds with such rhythms, and of failing to understand our fundamental need for the dialectics of time.*

We believe, however, that if human life is indeed placed in the framework of these natural rhythms, what we are determining is happiness, not thought. The mind needs a much closer pattern of reference points. If, as we would argue, intellectual life is to become the dominant form of life, physically speaking, with thought time prevailing over lived time, then we must devote all our efforts to the quest for an active repose that finds no satisfaction in what is freely bestowed by the hour and the season. It would seem that for Pinheiro dos Santos this active, vibrant *repose corresponds to the lyric state. The Brazilian philosopher has close knowledge of modern French literature, in particular of Valéry and Claudel, whom he greatly admires. He submits to each in turn, to the power and the rhetoric of Claudel's writing, and then to the subtle ambiguities of Paul Valéry's thought. In Valéry, he appreciates most of all the supreme art of the poet as, skillfully, he disturbs our calm and calms our disturbance, and moves from our heart to our mind, only to return at once from mind to heart.*

Yet Pinheiro dos Santos does not rest content with this rather coldly intellectual interpretation of the lyric life. He prefers that lyricism should continue to be regarded as a purely physical charm, a myth that lulls us to sleep, a complex binding us to our past, to our youth and its impetuosity. Indeed, he suggests a lyric myth for Rhythmanalysis which could well be called the Orpheus complex. This complex would correspond to our first and fundamental need to give pleasure and to offer solace; it would be revealed in the caresses of tender sympathy, and characterized by the attitude in which our being gains pleasure through the giving of pleasure, by the attitude of making some kind of offering. The Orpheus complex would be the exact antithesis of the Oedipus complex. Poetic interpretations of this Orpheus complex may be seen in Rilke's orphic lyricism, as

Félix Bertaux has called it, a lyricism which egotistically lives out an indeterminate love of others.[1] How very sweet it is to love anyone or anything, indiscriminately! How delightful ever to live at the moment of falling in love, ever amidst love's first rapturous declarations! This, then, is the basis of a theory of formal pleasure which is the very opposite of the theory of that material pleasure, immediate and objective, which in the Oedipus complex binds the unfortunate child to the face that he first sees above his cradle. Rhythmanalysis is the complete antithesis of Psychoanalysis in that it is a theory of childhood rediscovered, of childhood which remains a possibility for us always and opens a limitless future to our dreams. It is interesting to note here that Pinheiro dos Santos has, in an essay in which he takes issue with Freud's work on Leonardo, set out to explain the artist's creative genius in terms of an *eternal childhood*. Creationism is, in fact, nothing other than the process of growing perpetually younger, and a method of systematic wonderment which helps us rediscover a pair of wondering eyes with which to look upon familiar sights. Every lyric state must originate in this truly enthusiastic knowledge. The child is our master, as Pope once said.[2] Childhood is the source of all our rhythms and it is in childhood that these rhythms are creative and formative. The adult must be rhythmanalyzed in order that he may be restored to the discipline of that rhythmic activity to which he owes his own youth and its development.

We ourselves would prefer that the lyric state be subjected to some kind

---

1. Félix Bertaux was a Germanist and a contemporary of Bachelard's. He translated a number of German writers, including Ernst Wiechert and Thomas Mann, and produced French-German dictionaries. Bachelard probably refers here to Bertaux's book *Panorama de la littérature allemande contemporaine* (Paris: Kra, 1928).
2. This may well be an imperfect recollection on Bachelard's part, since this phrase is more reminiscent of Wordsworth's "The Child is Father of the Man," in his poem "My heart leaps up when I behold." If it is indeed a reference to Pope, then it is likely to be to "An Essay on Man," lines 275–82:

> Behold the child, by Nature's kindly law,
> Pleas'd with a rattle, tickled with a straw:
> Some livelier plaything gives his youth delight,
> A little louder, but as empty quite:
> Scarfs, garters, gold, amuse his riper stage;
> And beads and pray'r books are the toys of age:
> Pleas'd with this bauble still, as that before;
> 'Till tired he sleeps, and life's poor play is o'er!

of mental elaboration, thus setting ourselves at some distance from the unconscious and its powers, which imprison us in the Orpheus complex. We have, for this reason, turned our attention to the uppermost regions of superimposed time, to thought time, in our search for the most clear-cut and therefore most invigorating of dialectics.

We have, for example, sought to experience all Valéry's poetry in our own way by applying to these poems the structures implicit in the dialectics of time. This may well be too abstract and too personal an approach, suggested all too readily by habits bred of dry and dusty philosophy. Yet we have discovered that as a result of using this method, which in effect impoverishes, we can hear rare and precious echoes; we have discovered, in particular, that the temporal structure found in ambiguity can help us to intellectualize rhythms produced by sound, and so enable us to *think* that poetry which will not reveal all its charms when we confine ourselves to speaking or feeling the lines, and nothing more. We have come to realize that it is the idea that sings its song, that the complex interplay of ideas has its own particular tonality, a tonality that can call forth deep within us all a faint, soft murmuring. If we speak soundlessly and allow image to follow image, in quick succession, so that we are living at the meeting point, the point of superimposition, of all the different interpretations, then we shall understand the nature of a truly mental, truly intellectual lyric state. Reality will, in this way, be enfolded and adorned by the rich garment of conditionals. In place of the association of ideas there comes the ever possible dissociation of interpretations. The mind takes pleasure in its refusal of all that it once found unfailingly attractive: it discovers all the delights of poetry in its destruction of poetry, as it contradicts the sweet spring and resists all charming things. This, it must be said, is a highly epicurean asceticism, since in this conditional form pleasure would seem the more vibrant. Poetry is thus freed from the rule of habit, to become once again the model of rhythmic life and thought that it used to be, and so it offers us the best possible way of rhythmanalyzing our mental life, in order that the mind may regain its mastery of the dialectics of duration.

La Dialectique de la durée (1936; Paris: Presses Universitaires de France, 1950), 148–50.

Bachelard will offer a very different account of reading poetry only two years later when, in *La Psychanalyse du feu*, he describes it as "reverie," as communication between the reader's and the writer's unconscious. In the light of this, it would be overhasty to make *La Dialectique de la durée* the precursor of his work on poetry. Jacques Gagey has argued that the book in fact marks Bachelard's failure to achieve the "conversion" to poetry sought, he believes, since *L'Intuition de l'instant* in 1932, regarding it as evidence of a "crisis" caused by Bachelard's inability to find what Gagey calls "the basis for philosophical self-assertion" (1969: 79–81). This surely begs the question of Bachelard's "self," the "real" Bachelard being for Gagey the writer of the books on poetry. It may well seem somewhat illogical to interpret the written in terms of the unwritten, and untenable too, given that *La Dialectique de la durée*, far from being as remote from Bachelard's epistemology as it appeared initially, develops themes and ideas already noted in his previous work. For instance, his description of how he reads poetry, with its emphasis on thought, on the beneficial effect of ambiguity on the reader, and its complete—and surprising—neglect of poetic language, can be compared with his description in *Le Nouvel Esprit scientifique* of the effect on him of reading two chapters of Heisenberg's *Physical Principles of Quantum Theory*, the first one proving the validity of wave against particle theory, the second disproving this in favor of particle theory. This "dialectical approach" is described as bringing the reader "psychological benefit"; its paradoxes free him from realism and rationalism equally, and are regarded therefore as "excellent mental hygiene" (88). It may seem a far cry from Heisenberg to a poem by Valéry—Roch Smith discusses "Le Cimetière marin" in this context (1982: 68)—but for Bachelard microphysics and poetry have the same effect and teach the reader the same lesson: in both, to quote a phrase from the last extract, "reality will . . . be enfolded and adorned in the rich garment of conditionals." Poetry, like modern science, explores and exploits possibility. This, as I have shown, was already a theme in *Le Nouvel Esprit scientifique*, with its reference to "the poetic strivings of mathematicians" and the comparison drawn between mathematical symbols and Mallarmé's images. *La Dialectique de la durée* is best understood in this context, not of course in order to reduce it to what has gone before, but rather to see how what has gone before is extended, even deflected, here.

What is new is the idea of "vertical time." It is, as André Robinet has said, "a metaphor" (1974: 363), but a metaphor of what? Bachelard himself gives the source of this idea, an article by Alexandre Marc, "Le

Temps et la personne," which, together with his own article "Idéalisme discursif," appeared in *Recherches philosophiques 4* (1934–35). Marc—and Bachelard quotes him at length—argues that what is specifically human is not action, since animals share this characteristic, but the *"vertical* dimension," by which he means, he says, "the possibility of 'growing deeper,'" "the faculty of rupture and opposition" (*DD* 94–95). This is extraordinarily close to Bachelard's own thinking, and at the same time it helps him to clarify the paradoxical idea of the "deep subject" in *Le Nouvel Esprit scientifique*. For Marc as for Bachelard, depth and verticality are metaphors for human thinking. Thought is against life; its characteristics must therefore be opposite to those attributed to life, life being regarded as history, as continuity, simple, linear, and horizontal. A possible antonym for all this is "depth," which indeed Bachelard has already tried. Now, however, he is aware of the inappropriateness of this word to his purpose, given its association with psychoanalysis, which moreover makes depth and lived experience coincident. "Verticality" is clearly preferable, more in keeping with his ideas of progress and project.

Vertical time is by definition discontinuous, constructed by man, by his polemics with the world. It has its drawbacks, as all metaphors do, because, despite Bachelard's definition, a vertical is still a line, continuous, simple, self-sufficient. He counters this by repeatedly attaching to it words and ideas which establish its discontinuity—"superimposition," "tiers," "compound," "instantaneity," "hierarchy," "lacunae"—once again inflecting and "deforming" language in order to propose a nonlinear verticality. What attracts him to it is above all the notion of an "upward thrust," a separation and distancing from life. "We remain entirely convinced," he says, "that the mind thrusts far beyond the line of life," and this, more than discontinuity, is the heart of *La Dialectique de la durée*. This uncompromising opposition to the primacy of lived experience is surely the "philosophical self-assertion" Gagey was looking for. It shows Bachelard at his most polemical and his most subversive, against Western philosophy in general, where even for rationalism and idealism, he believes, the world is immediate and lived, and more particularly against prestigious contemporaries like Bergson, Freud, and the philosophers of existence. The new consciousness must break with life; its dynamic is what he calls "the axis of liberation." This, then, is the "psychic revolution" accompanying the "revolution in reason," this freeing of the mind from the "ideal of identification" (*Eng. rat.*, 12) imposed by life. *La Dialectique de la durée* is part of this revolution; it cannot in any way be seen as just failure to achieve poetry. Indeed, when Bachelard describes reading poetry as constructing difference, as escaping from linearity, we realize from the standpoint of more recent French philosophy how innovatory and subversive all this is.

Chapter 5

Psychoanalyzing Reason: 1938

> I believe that we learn *against something,* perhaps indeed *against someone,* and quite definitely, *against ourselves.*
> —*L'Engagement rationaliste*

La Formation de l'esprit scientifique: contribution à une psychanalyse de la connaissance objective, published in 1938, is probably Bachelard's best-known book in France, and the ideas of the "psychoanalysis of reason" and the "epistemological obstacle" it develops are familiar to anyone in that country who has, in recent years, studied philosophy at school or university.[1] The first extract I have chosen, "Epistemological Obstacles," from chapter 1 of the book, "The Idea of the Epistemological Obstacle," is well known for its exposition of these two ideas. The second, "The Teacher and the Taught," comes from the last chapter, "Scientific Objectivity and Psychoanalysis," and while it is less known, it is in many ways more revealing.

The year 1938 is something of a watershed in Bachelard's work. For the previous ten years, he was chiefly concerned with the "new scientific mind," the epistemology of post-Einsteinian science, and although his ideas and values may at times have appeared to follow another course, this, as we have seen, was still directed by his epistemology. After *La Dialectique de la durée*, which with its "psychology of annihilation" maintained this direction to a surprising degree, despite initial impressions, Bachelard returned in his next book to epistemology, examining in *L'Expérience de l'espace dans la physique contemporaine* (1937) the impact of Heisenberg's uncertainty principle on our conception of reality. The epistemological break is especially striking in the context of probability theory, and Bachelard drew attention to this by describing the "objects" of microphysics as "experimental moments," "the different and successive states of probable becoming" (34). Individuality and identity do not characterize fundamental particles, and consequently we should no longer set any store by them: they may well be the traditional ways of describing and determining what is "real," but modern science has proved them invalid. In retrospect, the years 1936 and 1937 saw Bachelard pursuing difference in both subject and object, as indeed he had done since 1928. But in 1938, with *La Formation de l'esprit scientifique* and also *La Psychanalyse du feu* published later that year, his ideas take another course, leading quite unexpectedly from science to poetry. Something of this change is suggested by the word "formation" which has the sense not just of "forming" but of "structure." Whereas in *Le Nouvel Esprit scientifique*, Bachelard was concerned with "forming" the mind in the pedagogical sense, with "opening" it in accordance with the new science, making it "project," now all of a sudden he appears fascinated by the mind as a structure formed by the past, and this fascination will eventually turn him from science toward poetry, by way of a new conception of consciousness.[2]

How and why this change comes about has been much discussed, Bachelard's commentators being, as so often, at loggerheads over the implications of this "psychoanalysis of objective knowledge." Jean-Pierre Roy applauds it for "de-ideologizing" science, purging it of imagination (1977: 29–31). Lecourt and Vadée, on the other hand, see in it confirmation of Bachelard's "ideology," Lecourt arguing that it proves the "epistemological illusion" which, in the end, denies Bachelard's materialist and dialectical philosophical theses (1974: 134), Vadée repeating that Bachelard totally misunderstands "the real basis of science in the necessity of the development of material production and of mastery of nature (physical and social)" (1975: 62). Bachelard, in Vadée's opinion, simply "flirts" with psychoanalysis (61), but Jacques Gagey considers his psychoanalysis as "very—perhaps too—classical" (1969: 93), as a step taken toward imagination, toward life, toward what he regards as the "real" Bachelard (100–102). This dissension tells us rather more about the commentators than about Bachelard, and underlines the need to return to the text, to pay close attention to what Bachelard himself wrote.

Bachelard states his intention very clearly in the first of these two extracts. It is to rid the mind of all the "causes of inertia" that impede progress. This indeed has been his aim since 1933, when in *Les Intuitions atomistiques* he referred to his "task of catharsis." What is new and important in *La Formation de l'esprit scientifique* is the idea of "epistemological obstacles," the idea that what obstructs knowledge is not just resistant matter, not just the shortcomings of reason, but something in the act of cognition itself. Knowledge is won *against* previous knowledge—familiar ground again—but these past errors are now understood as other than purely scientific, as in fact psychological. The "three states" through which every scientist must progress—the concrete, the concrete-abstract, the abstract—may coexist, so that even the most mathematical of modern scientists can yield to "naïve curiosity" and "wonderment" (7–9). The scientist's mind is not, Bachelard argues, *tabula rasa,* it is thoroughly prejudiced, marked by preconceived ideas and values. Any history of scientific ideas will provide abundant examples of these epistemological obstacles, as will the classroom experience of a science teacher (Bachelard writes with feeling here!). History, teaching, and even everyday life ought to alert us to the fact that the same word, denoting the same experience, the same idea, means different things to different people, and that this difference has a psychological, affective basis. The historian must beware of this, the epistemologist and the teacher must use it, so that scientists and children can learn to think more effectively, more dynamically. The "psychoanalysis of reason," as this first passage so clearly shows, shares the aim of all Bachelard's work; this "intellectual and emotional catharsis," this "cognito-affective control" seeks to "open" reason, to allow it to progress.

Why then speak of a change of direction in this book? There might seem to be nothing new in all this, apart from Bachelard's references to affectivity. Yet this element is very new, and it is this, as the second extract suggests, that leads Bachelard to change direction, or better, to change his values in the course of this book. He becomes less concerned with forming the scientific mind, and more and more absorbed in what he discovers of its formation. Most of *La Formation de l'esprit scientifique* is devoted to an examination of epistemological obstacles in what Bachelard calls the "prescientific mind," observed through the writings of sixteenth-, seventeenth-, and eighteenth-century scientists and alchemists. Chapter by chapter, he lists these obstacles: first (ideas familiar to us from the notion of the epistemological break), immediate experience, the high value placed on generality in knowledge, on the unitary and the successful; next (something new), ideas about the mystery of substance and of living things; then finally the most powerful epistemological obstacle of all, the libido, sexuality. Some two hundred pages explore these obstacles and provide us with a wealth of examples. We are fascinated, as Bachelard so plainly is, by what they reveal of the human mind, so that far from accomplishing his aim of ridding the mind of these "causes of inertia," of "hunting down sin" as Descombes puts it (1980: 91), Bachelard keeps them before our attention. He becomes, to quote Jacques Gagey, the "accomplice of the libido he claims to denounce" (1969: 96). He now appears to accept what he previously—and vigorously—refused, the power of the unconscious in us. References to Freud, Karl Abraham, Havelock Ellis, Otto Rank, and most frequently of all, Ernest Jones, show that Bachelard has a much better knowledge of psychoanalytical theory than before. However, it is the evidence of his own reading of prescientific texts, of the alchemists in particular, that leads to his discovery of the new dimension in man's being, "the heavy burden of ancestrality and unconsciousness" (209). As this second extract makes clear, this instinctual, affective part of us is not just an obstacle; it can be a spur to knowledge. It is therefore far more than a question of something that is hard to eliminate because it is so pervasive; it is in fact something that must not at all costs be eliminated. Hence the moral tone of this book that Michel Serres complains of (1974: 73), for Bachelard is excited by what he has discovered so unexpectedly about human nature.

At the end of *La Formation de l'esprit scientifique*, Bachelard turns his attention from alchemy back to the classroom, from a remote and alien experience, to the mathematics and science lessons we have all lived through, analyzing what he calls the "rational emotion." He considers this first in relationships between pupil and teacher, between one pupil and another, then second in the "lone scientist." This awareness of the "*social* dimension in rational convictions" is also new in Bachelard, doubtless brought to his attention by psychoanalysis with its emphasis on the importance of interpersonal relationships.[3] He now accepts that emotions and instincts are involved in the pursuit of knowledge, that if a child is to make progress, he must feel he is right *against* someone, against someone else's errors. Being right is a feeling, not just a mathematical conviction. Correcting mistakes is more than a matter of rectification, for as Bachelard shows, we can have a number of different psychological attitudes to error. The polemic of reason and reality, and the dialectics of rationalist and empiricist, have an emotional charge, without which they would in fact cease to be. Bachelard discovers in human beings the will to be rational, not, as his language suggests, out of a Nietzschean will to power over others, but rather out of a Freudian desire for *satisfaction,* for rational satisfaction through others. Where the "lone scientist" is concerned, a psychoanalysis of reason will reveal, Bachelard believes, the importance of subjective rectification, of psychological modification: if intellectual creativity is accompanied by consciousness of that creativity, then the pleasure derived from this triumph over past errors will provide the impetus to even greater creativity.

Extract I
Epistemological Obstacles

When we start looking for the psychological conditions in which scientific progress is made, we are very soon convinced that *the problem of scientific knowledge must be posed in terms of obstacles*. We are not talking about external obstacles, such as the complexity and the transience of phenomena, nor indeed are we laying the blame upon the weakness of our senses or of the human mind: it is at the very heart of the act of knowing that, by some kind of functional necessity, sluggishness and disturbance arise. It is here that we shall see the causes of stagnation and even regression, and here, too, that we shall be able to discern the causes of inertia that we shall term epistemological obstacles. Our knowledge of reality is like a light which always casts a shadow in some nook or cranny. It is never immediate, never complete. Reality's revelations are always recurrent. Reality is never "as we might imagine it" but rather always as we ought in fact to have thought. Empirical thought is clear *after the event*, when all the apparatus of reason has done its work. Whenever we look back and see the errors of our past, we discover truth through intellectual repentance. Indeed, we know *against* previous knowledge, when we destroy knowledge that is imperfect and surmount all those obstacles to spiritualization that lie in the mind itself.

The notion that we can go back to the very beginning and start from nothing where the creation or increase of our possessions is concerned can occur only in a cultural system based on juxtaposition alone, in which something that is known is immediately something that enriches. Yet when our soul confronts all the mystery of reality, it cannot make itself ingenuous, simply by decree. It is quite impossible, then, to erase every single trace of our ordinary, everyday knowledge once and for all. When we contemplate reality, what we think we know very well will cast its shadow over what we ought to know. Even when it first approaches the cultural domain of science, the mind is never young. It is, in fact, rather old, as old as its prejudices. When we enter the realms of science, we grow intellectually younger, and we submit to a sudden, complete mutation that must contradict the past . . .

The concept of the *epistemological obstacle* can be examined in the historical development of scientific thought and also in educational practice. In both these areas, such an examination will prove far from easy. History is, in fact, intrinsically hostile to all normative judgments. We are obliged, however, to take a normative view if we wish to evaluate the efficacy of thought. Nothing we find in the history of scientific thought is of any possible use in contributing to the evolution of that thought. There are some kinds of knowledge which, even though they are accurate, will bring useful research to a premature end. The epistemologist must be selective, then, in his use of the material historians provide. He has to evaluate these documents from the point of view of reason, or rather from the point of view of evolved reason, for it is only now that we can really judge the errors of the intellectual past. Moreover, even in experimental science, it is always rational interpretation that establishes facts in their correct position. Success and danger both lie along the axis that joins experience and reason, and in the direction of rationalization. Reason alone can dynamize research, for it is reason alone that goes beyond ordinary experience (immediate and specious) and suggests scientific experience (indirect and fruitful). The epistemologist must pay close attention, then, to this striving toward rationality, toward construction. We see here what it is that distinguishes the epistemologist's calling from that of the historian. The historian of science has to take ideas as facts. The epistemologist has to take facts as ideas, and place them within a system of thought. A fact which a whole era has misunderstood remains a *fact* in the historian's eyes. For the epistemologist, however, it is an *obstacle*, a counterthought.

It is when we examine the concept of the epistemological obstacle in greater depth that we can best discern the true intellectual value of the history of scientific thought. Although the preoccupation with objectivity leads the historian of science to catalogue his texts in great detail, all too often it fails to take him further, that is, to the measurement of psychological variations in the interpretation of just one text. The same word can, at the same period in time, have within it very many different concepts. What misleads us here is the fact that it is the selfsame word that both indicates and explains. What is indicated stays the same, but the explanation changes. The telephone, for instance, is understood in very different ways by the subscriber, the operator, the engineer, and the mathematician concerned with the differential equations of the telephone current. The epistemologist must make every effort to set scientific concepts within effective psychological syntheses, that is to say, within progressive psychological syntheses, by establishing an array of concepts for every individual idea, and by showing how one concept produces another and how it is related to another. Then, perhaps, he may succeed in measuring epistemological efficacy. Straightaway, scientific thought will be seen as a difficulty that has been overcome, an obstacle that has been surmounted.

Turning now to consider education, we see that the concept of the pedagogical obstacle is equally ill understood. I have often been struck by the fact that science teachers, even more than other teachers, if this is at all possible, cannot understand that their pupils may not understand. Very few of them have made a close study of the psychology of error, of ignorance, and of heedlessness. Gérard-Varet's book has met with little response.[1] Science teachers imagine the thinking mind to proceed like a lesson; they imagine, too, that by repeating a class we can always make good the slapdash knowledge we have indifferently acquired, and that we can be made to understand a proof by having to chant it parrot fashion. They have not given any thought to the fact that when the young person comes along to the physics lesson, he already possesses a body of empirical knowledge: it is not, therefore, a question of *acquiring* experimental culture but rather of *changing* from one experimental culture to another, and of removing the abundance of obstacles that ordinary life has already set up. Let us take just one example here: the buoyancy of a floating body is the object of an everyday intuition that is shot through with errors. The activity here is ascribed, more or less openly, to the floating body, or rather to the *swimming* body. If we put our hands on a piece of wood and try to sink it, it will resist. We find it hard to ascribe this resistance to the water. It is not easy, therefore, to teach Archimedes' principle so that it is understood in all its marvelous mathematical simplicity, if we have not first criticized and de-

---

1. Bachelard's footnote: Louis Gérard-Varet, *L'Ignorance et l'irréflexion: essai de psychologie objective* (Paris: Alcan, 1899). Note amended and corrected.

stroyed this complex and impure body of primary intuitions. In particular, without such a psychoanalysis of initial errors, we shall never be able to teach children to see that the body that emerges from the fluid and the body that is completely immersed both obey the very same law.

All scientific culture must begin, then, as we shall later explain at some length, with an intellectual and emotional catharsis. The hardest of our tasks still remains: we must put scientific culture on the alert, so that it is always ready to move; we must replace closed, static knowledge with knowledge that is open and dynamic; we must dialectize all experimental variables; and last, to reason we must give reasons for developing.

We could, moreover, generalize these observations: they are at their most apparent in the teaching of science, but they are relevant to all aspects of education. In the course of a career that has been long and varied, I have never seen a teacher change his teaching methods. A teacher has no *sense of failure* precisely because he considers himself a master. He who teaches commands. Hence a great flood of instincts. Von Monakow and Mourgue have noted the problems experienced in attempts to reform teaching methods, drawing attention to the part played here by the mass of instincts in every teacher.[2] "There are individuals for whom any advice with respect to the *educational errors* they commit is completely and utterly useless, because these so-called errors are simply the expression of instinctive behavior." Von Monakow and Mourgue are, of course, discussing "psychopathic individuals," yet the psychological relationship between master and pupil is a relationship that can easily become pathogenic. The teacher and the taught are both dependent on a special kind of psychoanalysis. We must not, in any case, neglect the study of the lower forms of the psyche if we wish to describe every aspect of mental energy, and so prepare that cognito-affective control that is indispensable for the progress of the scientific mind. More precisely still, it is by revealing epistemological obstacles that we can help to found the rudiments of a psychoanalysis of reason.

La Formation de l'esprit scientifique: contribution à une psychanalyse de la connaissance objective (1938; Paris: Vrin, 1972), 13–14, 17–19.

---

2. Bachelard's footnote: Constantin von Monakow and R. Mourgue, *Introduction biologique à l'étude de la neurologie et de la psychopathologie: intégration et désintégration de la fonction* (Paris: Alcan, 1928), 28. Note amended. This book has not been translated; my translations here.

Extract II
The Teacher and the Taught

We must take care not to exaggerate the influence of formal education. Indeed, as Von Monakow and Mourgue observe, relationships with the peer group are for the schoolchild far more formative than relationships with older people, and classmates are much more important than teachers. Teachers provide ephemeral, haphazard knowledge, especially in secondary schools with all their incoherent multiplicity, knowledge which also bears the pernicious stamp of authority. On the contrary, our schoolmates implant indestructible instincts in us. We ought, therefore, to take a group of pupils and encourage them toward an awareness of a group reason; in other words, we should help them to acquire the instinct toward social objectivity, for this is an instinct that is much underestimated and in whose place we prefer to develop the opposite instinct of *originality*, failing to see that this originality which we learn from our literary studies is contrived and artificial. To put it another way, if objective science is to be really educational, then the way it is taught must be socially active. Normal educational practice makes a very big mistake when it establishes an inflexible relationship between master and pupil. In our view, the fundamental principle of the *pedagogics* of the objective attitude is this: *whoever is taught must teach*. Any teaching that is received and not then passed on to others will produce a mind entirely devoid of dynamism and self-criticism. In science subjects in particular, this kind of teaching makes knowledge fixed and dogmatic, whereas this very knowledge ought to spur the pupil on to further progress and invention. Most important of all, it fails to provide the psychological experience of human error. I can imagine only one admissible use for school "compositions" and that is as a means of choosing monitors, who would pass on a whole range of lessons, gradually decreasing in rigor. The pupil who is top of the class would be rewarded by the pleasure of teaching the second to top, the second the third, and so it would continue until the point where errors really are far too important and pervasive. The tail end of the class is not, however, without its usefulness from the psychologist's point of view: it exemplifies the nonscientific species, the subjectivist species, whose immutability is highly instructive. We can forgive the rather inhuman way in which the dunce is used in many a mathematics lesson if we remember that he who is wrong objectively considers himself right subjectively. Members of the cultured middle classes think it rather smart to boast of their total ignorance where mathematics is concerned. People will really wallow in their failure, once that failure is sufficiently plain. Furthermore, the existence of a group which is entirely immune to scientific knowledge is most helpful with regard to our attempts to psychoanalyze rational convictions. It is not enough for a man to be right, but he must be right *against* someone. If this *social* dimension is lacking in rational conviction, then our sense of being profoundly in the right is not far from being a feeling of resentment; any conviction that is not put to the test by our efforts to teach it to someone else will act in our soul like an unrequited love. Indeed, what proves the psychological healthiness of modern science as compared with eighteenth-century science is that there is now a steady decrease in the number of things that are *misunderstood* . . .

If we have allowed ourselves to describe, though briefly, this utopia of the schoolroom, it is because it seems to offer us, proportionately speaking, a practical and tangible way of measuring the psychological duality of rational and empirical attitudes. We believe, in fact, that in any real teaching there is an interplay of philosophical nuances: *the teaching we receive is, psychologically speaking, a kind of empiricism; the teaching we give is, psychologically speaking, a kind of rationalism*. I listen to you: I am all ears. I talk to you: I am all mind. Even if we are both saying the same thing, what you say is always somewhat irrational; what I say is always somewhat rational. You are always slightly in the wrong, and I am always slightly in the right. What is being taught is of little importance. It is the psychological *attitude*, composed of resistance and incomprehension on the one hand, and on the other, of impulse and authority, that comes to be the decisive factor in any real teaching, when books are left behind and we talk, instead, to men.

Now, since objective knowledge is never complete and since new *objects* never cease to provide new topics of conversation in the dialogue between the mind and things, any real teaching of science will be drawn this way and that by the ebb and flow of empiricism and rationalism. Indeed, the history of scientific knowledge is an endlessly renewed alternation of empiricism and rationalism. This alternation is more than just a fact. It is a necessity for our psychological dynamism. It is for this reason that any philosophy that confines culture to realism or nominalism only sets up the most formidable of obstacles for the development of scientific thought.

In order to cast light on the interminable argument between rationalism and empiricism, Lalande has suggested, in a splendid extemporization at a recent congress of philosophers, that a systematic study be made of the periods in which reason finds great satisfaction and of those in which reason is ill at ease. He showed that in the course of the development of science, all of a sudden syntheses occur that seem to swallow up empiricism, like for instance Newton's synthesis of mechanics and astronomy, Fresnel's of vibration and light, and Maxwell's of optics and electricity. At such times, the teacher is triumphant. Yet the brightness will fade and darkness gather; something is going wrong, for Mercury stirs in the heavens, photoelectric phenomena fragment the wave, and fields cannot be quantified. At such times, doubters are wreathed in smiles, like schoolboys. Were we to extend the enquiry Lalande has suggested, we should be able to determine very precisely what exactly is meant by the *satisfaction* of reason when it rationalizes a fact. We should then see, with as much accuracy as possible, and with respect to specific instances in the secure domain of past history, the passage from the assertoric to the apodictic, and the illustration of the apodictic by the assertoric.

Yet while this purely historical enquiry will provide us with the quasi-logical meaning of the satisfaction of reason, it cannot offer us the psychology of *the feeling of being in the right*, in all its complexity and in the ambivalence of gentleness and authority. If we are to know all the emotions involved in the use of reason, we shall have to live and also teach a scientific culture, defending it against all irony and incomprehension, and then arm ourselves with it, and sally forth against philosophers, against psychologists of the inner life, pragmatist and realist alike. Then shall we have some idea of the range of values associated with the rational emotion: when it is men who put us in the right with regard to other men, we taste the sweet success beloved of the politician with his will to power. When, however, it is things that put us in the right with regard to other men, we witness the triumphant success not of the will to power, but of the will to be rational, in all its brilliancy, *der Wille zur Vernunft*.

Yet things cannot put the mind in the right once and for all. There is no doubt either that this rational satisfaction must be renewed if it is to provide a real psychic dynamism. Through a curious effect of habit, the apodictic, now grown old, acquires a taste for the assertoric, and the *rational fact* remains, without the rational apparatus. The only thing that men have grasped in the whole of Newton's mechanics is that it is the study of gravitation, whereas for Newton himself, gravitation was a metaphor, not a fact. We have all forgotten that Newtonian mechanics assimilated apodictically the *parabola* of the movement of projectiles on earth with the *ellipse* of planetary orbits, and this by means of a rational system. We must therefore take steps to prevent rational truths from degenerating, for they tend to lose their apodicticity and deteriorate into intellectual habits. Balzac said that bachelors and old maids put habit in the place of emotion. In exactly the same way, teachers put lessons in the place of discovery. Teaching about the discoveries that have been made throughout the history of science is an excellent way of combating that intellectual sloth which will slowly stifle our sense of mental and spiritual newness. If children are to learn to invent, it is desirable that they should be given the feeling that they themselves could have made discoveries.

We must also disrupt the habits of objective knowledge and make reason uneasy. Indeed, this is part of normal pedagogical practice. It is not without a touch of sadism, which shows us fairly clearly the presence of the will to power in the science teacher. This teasing use of reason operates in the reverse direction, too. In our ordinary day-to-day lives, we love putting someone else in a spot. The person who asks riddles provides us with a most revealing example here. Often, a riddle that comes out of the blue is the revenge of the weak against the strong, of pupil against teacher. When a child asks his father a riddle, in all the ambiguous innocence of intellectual activity, is he not satisfying his Oedipus complex? And vice versa, it is not hard to psychoanalyze the attitude of the mathematics teacher, serious and awesome as the sphinx.

Last, we can discern in certain educated minds a real intellectual masochism. They need some kind of mystery behind the clearest solutions in science. They are reluctant to accept the clear, self-conscious evidence furnished by axiomatic thought. Even when they have conquered and mastered a mathematical concept, they still need to postulate some kind of realism that lies beyond their grasp, crushing them. In the physical sciences, they postulate reality's fundamental irrationality, although, with respect to the phenomena of the laboratory, thoroughly mastered and mathematized as these phenomena are, this irrationalism is but the *result of all the carelessness* perpetrated by the experimenter. Yet the mind does not seek to enjoy, in peace and quiet, knowledge that is completely closed in on itself. It thinks not of present difficulties but of those of tomorrow; it thinks not of the phenomenon securely imprisoned in the apparatus now in use, but rather of the phenomenon that is free and untamed, impure and hardly even named! Philosophers turn this unnamed thing into the unnameable. Brunschvicg recognized that this duality, marked as it is by contrary values, is present even in the foundations of arithmetic, for he spoke of a science of number which is used either to prove or to impress and dazzle, meaning of course that before we dazzle others, we must first blind ourselves.[1]

These sadistic and masochistic tendencies, which are found in the social life of science in particular, do not however provide an adequate description of the real attitude of the lone scientist; they are no more than the first obstacles that the scientist must surmount if he is to acquire complete scientific objectivity. In the present state of scientific development, the scientist faces the continuing need to *renounce his own intellectuality*. If there is no explicit renunciation, and no relinquishment of intuition, no abrogation of favorite images, then objective research will soon lose not just its fruitfulness but the very vector of discovery, the inductive impetus. We must constantly strive toward desubjectivation if we are to live and relive the instant of objectivity, if we are to remain forever in the *nascent state* of objectivation. The mind that psychoanalysis has freed from the twofold slavery of subject and object can savor the heady delight of oscillating between extraversion and introversion. An objective discovery is at once a subjective rectification. If the object teaches me, then it modifies me. I ask that the chief benefit the object brings should be an intellectual modification. Once pragmatism has been successfully psychoanalyzed, I wish to know for the sake of knowing, never for the sake of using. Conversely, too, if through my own efforts I have been able to obtain some psychological modification—which we can only imagine as a complication on the mathematical level—then armed with this essential modification, I go back to the object, I call upon experiment and technique to illustrate and bring about the modification already brought into being psychologically. Certainly, the world will often resist, the world will always resist, and the efforts of mathematics must be ever renewed, growing ever more flexible, and constantly rectified. They are rectified however by their very enrichment. Suddenly, these efforts at mathematization are so successful that reality crystallizes along the axes provided by human thought, and new phenomena are produced. Indeed, we can talk with complete confidence now of the creation of phenomena by man. The electron existed before twentieth-century man. Yet before twentieth-century man, the electron

---

1. Bachelard's footnote: Léon Brunschvicg, *Le Rôle du pythagorisme dans l'évolution des idées* (Paris: Hermann, 1937), 6.

did not sing. Now in the triode valve the electron sings. This phenomenological *realization* occurred at a precise point in mathematical and technical development, the point at which it came to maturity. Any attempt at a premature realization would have been in vain. Had astronomy sought to *realize* the music of the spheres, it would have failed. It was but a meager dream that gave a meager science value. The music of the electron in an alternating field has, on the other hand, proved to be realizable. This dumb being has given us the telephone. This same invisible being will give us television. Man triumphs, then, over the contradictions of immediate knowledge. He forces contradictory qualities to become consubstantiate, as soon as he has freed himself of the myth of substantialization. There can be no irrationalism in a substance that organic chemistry has made with great care and attention: irrationalism could only be an impurity. Such an impurity can, moreover, be tolerated. The moment it is tolerated, we see that it is quite powerless, and in no way dangerous. Functionally speaking, this impurity does not exist. Functionally speaking, a substance realized by modern chemical synthesis is entirely rational.

La Formation de l'esprit scientifique: contribution à une psychanalyse de la connaissance objective (1938; Paris: Vrin, 1972), 244–49.

In *La Formation de l'esprit scientifique*, Bachelard's debt to psychoanalysis is plain, yet it is also plain that there is still much that he does not accept. Even though he is fascinated by the sexual dimension of alchemy, he does not in the end ascribe any role to sexuality, to the unconscious in man's development. His former values persist: it is still, as he declared in *L'Intuition de l'instant* (1932), "thought that rules our being," and consciousness that constitutes our truth (71). Yet now, with the help of psychoanalytical theory, Bachelard perceives that scientific thought has an essential affective basis, that consciousness is, as a result, impure. The subject is transcended not just by objects but by other subjects. He is defined by the emotions they arouse, and given depth by this interplay of psychological attitudes. The psychoanalysis of reason reveals how hard it is for human subjects to sustain their project, to overcome the epistemological obstacles that impede their dynamism, yet at the same time how necessary consciousness of having overcome those obstacles is to their willingness to pursue that project.

Chapter 6

Dynamic and Material Imagination: 1938–1948

> It is our belief that man takes *all* that he learns from the external world, and that he understands himself in terms of the pure kinematics of surrounding Nature.
> —*Essai sur la connaissance approchée*

There is a pleasing symmetry in Bachelard's work—ten years' publishing on science followed by ten on poetry—but it is not one that he planned, and far from being, in the eyes of his interpreters, something they can simply admire, it has become a stumbling block. The two aspects of his work are proved by some to be quite distinct, by others to have unity; his work on poetic imagination is for certain critics a betrayal of his epistemology, while others regard it as paramount, the crown of that epistemology.[1] To take sides at this stage would be inappropriate, but one point must be made, namely that whatever their view, the proponents destroy the peculiarly Bachelardian tension between science and poetry, reason and imagination. In later years, in his last three books on poetry especially (1957–61), Bachelard will declare the two aspects of his work to be different. Difference for him does not, however, imply contradiction or total opposition. We must therefore take care that our mental reflexes do not simplify and reduce it in this way.

Bachelard insists on the "clear polarity of intellect and imagination" (*PR* 45), opposite poles which exclude rather than attract each other (*PR* 47). At the same time, he also makes it clear that these are the poles of human consciousness, of what he calls a "working consciousness," at its best when alternating between images and concepts (*PR* 47). The dualism that J.-P. Roy so severely censures in Bachelard is not, as he would have it, the dualism of subjectivity and objectivity, life and thought, humanism and science, of contradictory methods and values inadmissibly coexisting

(1977: 11–66, 207–20). Roy remains curiously unaware that for Bachelard, science is a human activity, that his examination of objects and objectivity in science not only complicates our accepted idea of them but has as its clear corollary a conception of the scientific subject. Moreover, from the beginning, the poetic images that interest Bachelard are to do with objects; they are, as he puts it in *La Psychanalyse du feu* (1938), "centered" on objects (32). He describes reverie—a synonym for imagination here—as "objectively specific" (33), and says that the aim of this book is to determine "the objective conditions of reverie" (179). The polarity of scientific reason and poetic imagination is plainly not the simple opposition of objectivity and subjectivity. "Poetry," he writes in an article, "Instant poétique et instant métaphysique' (1939), "should give us both a view of the world and the secret of a soul, a being and objects at one and the same time" (*DR* 224).[2] If science and poetry are different, it is because in each there is a different relationship, a different tension between subject and object. This brings us back to what I have called Bachelard's subversive humanism, to his desire to abolish the frontiers of the internal and external worlds. Alternating between the poles of concept and image, we experience the changing tensions of subject and object, the shifting and breaching of their familiar frontiers, the "unfixing" that ensures what he will call very simply a "good consciousness" (*PR* 47).

Science and poetry are, for Bachelard, not just different activities, but different human activities, this one extra and essential word changing the sense of "difference." Both activities coexist in his own life, and overlap: *La Psychanalyse du feu* (1938), his first book on poetry, grows out of *La Formation de l'esprit scientifique; La Philosophie du non*, one of his most important epistemological works, is published in 1940, a year after his *Lautréamont*; the first chapter of *Le Rationalisme appliqué* (1949) appears as an article in 1947, a year before the two books on images of the earth. Most remarkable of all, the four seminal books on poetic images, *L'Eau et les rêves* (1942), *L'Air et les songes* (1943), *La Terre et les rêveries de la volonté* (1948), *La Terre et les rêveries du repos* (1948), are the work of France's leading philosopher of science, the professor of the history and philosophy of science at the Sorbonne. In addition, three further books on science are published between 1949 and 1953, followed in the years 1957 to 1961 by three more on poetry. Bachelard's lectures at the Sorbonne were, by all accounts, unusual for a philosopher of science. René Poirier and Jean Lescure have both described them in vivid terms (1974: 10–14, 227–28),

Lescure continuing his description in *Un Été avec Bachelard*, the controversial book published and republished, with expunctions, in 1983.[3] Georges Jean in *Bachelard: l'enfance et la pédagogie* (1983) also records his impressions of lectures attended in 1942 and 1943. Bachelard lectured separately on both epistemology and poetry, he points out, describing how he and others would sit through lectures on epistemology in order to be sure of a place at those on poetry (11–13). The fact that Bachelard exceeded his "job description" in this way is surely important for a better understanding of the duality of his interests, which in his own working life did not conflict or give rise to demarcation disputes. Both Jean and Lescure describe how Bachelard's lectures were assiduously followed by a number of poets, Lescure adding that as a result Bachelard came to spend time socially with these young men, taking up and developing in his lectures points first made in mealtime discussions (1983: 24, 1974: 228). Poetry is obviously at this time very much a part of his life, and even before this, we know that he taught literature to foreign students at Dijon.[4] Indeed, as Vincent Therrien has shown, Bachelard had from early childhood loved reading and excelled in literary studies at school (1970: 44). Yet this does not really explain the value poetry came to have for Bachelard. He was not a dilettante, who relaxed by reading and writing about poetry: he wrote nine books on poetry, books which according to Roland Barthes led to the founding of "a whole critical school" (1963: 739),[5] and which Gilbert Durand believes to have obliged French thinkers for the first time to take imagination seriously (1969: 15–27, 31–32, 45–46).

How then does Bachelard come to write about poetic images? He has referred to poetry in earlier books, but despite the interest and significance of these references, he was not, as we have seen, concerned with poetry in itself. The redirection of his work was in fact unplanned; the turning point seems to have been *La Psychanalyse du feu*, published in 1938 and intended to be a companion to *La Formation de l'esprit scientifique*, to "illustrate its general argument" by means of the example of fire which, with all the immediate and subjective values it accumulates, is an obstacle to scientific knowledge that must be "psychoanalyzed," that is to say, eliminated (*PF* 14–15). However, Bachelard discusses fire as an epistemological obstacle only in the fourth and fifth chapters, the first three exploring the values attributed to fire in his own childhood memories, in mythology, anthropology, and above all in poetry, the last three developing a theory of poetic imagination that will deflect his work for the next decade. Bachelard has described *La Psychanalyse du feu* as "both disorderly and incomplete . . . a book I would like to rewrite" (Christofides 1962: 267). Luckily for us, he did not rewrite it, for though the book is certainly disorderly, countering his initial position that scientific truth is the one and only truth with the discovery that there is such a thing as poetic truth, and then proposing contradictory versions of this new truth, this very disorder reveals something of the shock of Bachelard's discovery and the originality of the ideas he is trying to formulate. He discovers, *against himself,* that fire is not just an epistemological obstacle, that the overvaluation of fire, universal and enduring as it so clearly is, reveals something of great importance for our understanding of human beings. To explain this, Bachelard puts forward the hypothesis—converted in due course into the *"law of the four elements"* (*ER* 4)—that the dreams of all of us and the images of every poet are determined by an affinity with earth, air, fire, or water, an unconscious affinity which persists despite the evidence of both reason and experience. "I realized," Bachelard will say some years later, "that this kind of obsession with the four elements corresponded to some sort of human necessity" (Christofides 1962: 267). Fanciful it may seem, but this "tetravalence of reverie" (*PF* 148) is Bachelard's first version of poetic truth, the "true," "sincere" poet, as he puts it (*PF* 148), being faithful in his images and his language to one of the four elements (*PF* 148–51). This will, moreover, provide the framework of the four books published between 1942 and 1948 on images of water, air, and earth. Yet there is a problem here, for although the four elements produce dynamic, material images which can be related to his earlier work, the notion of unconscious affinity between man and matter conflicts with it.

The Freudian view of the unconscious is not, however, shared by Bachelard, despite initial ambivalence and disorderly thinking. He modifies Freudian theory, and as the extracts in this chapter and in Chapter 7 will demonstrate in neither his theory nor his practice of poetry is Bachelard a psychoanalytical critic. He differs with Freud on three main points. First, it is not interpersonal relationships with their instinctual, sexual basis that are formative for human beings, but their relationship with matter. Second, this relationship with matter is intellectually rather than sexually charged, the fundamental drive being what he calls the *"will to intellectuality"* (*PF* 26). Third, as a result of this, poetic images of matter do not spring from our instinctual depths, but instead arise in the "intermediate zone" between the unconscious and the rational consciousness, at the threshold of rational thought, of objective knowledge about the world (*PF* 26, 32).⁶

Bachelard sums up his position in one succinct phrase: poetic images belong to "the zone of material reveries that precede contemplation" (*ER* 6). The word "reveries" here points up one of the problems he faces, the problem of finding words to express his conception of imagination. He distinguishes between *rêve* and *rêverie*, making *rêverie* synonymous with poetic imagination. Reverie is not mere daydreaming; it is more the free play of the mind around objects, "centered" on objects, unlike *le rêve*, the night dream, pure subjectivity, unconsciousness of the world (*PF* 32). Bachelard will continue to make this distinction, perhaps most clearly in *La Poétique de la rêverie* (1961) when he writes that in reverie there is always a "glimmer of consciousness," always the presence of the dreamer (129). So far, so good, but when it comes to expressing the difference between the activities and the agents, then verbs—and nouns—fail him: *rêver* and *rêveur*, "to dream" and "dreamer"—have to serve in both contexts. Hence what may appear to be conceptual confusion in fact comes down to an absence of vocabulary. Bachelard tries to avoid these pitfalls by, for instance, using "oneiric" as the adjective more appropriate to "reverie" than "unconscious," but as these extracts show, there is some unevenness. He does use the word "unconscious," for example, though not in the strictly psychoanalytical sense, and readers should be aware of these linguistic problems, which sometimes result in words not meaning what we too hastily think they mean. Material imagination is, as Sartre put it when discussing Bachelard in *L'Être et le néant*, "a real discovery" (1943: 690 [1957: 600]), and its very newness requires and then rewards patient reading.

Bachelard's polemic with Freud forces him to admit, *against himself,* that it is not just "thought that rules our being" (*II* 71), that science is only one aspect of our relationship with matter. Freud also helps him to understand, *against Bergson,* that "man is created by desire, not by need" (*PF* 34), by what Bachelard interprets as his desire to know, rather than, as Bergson argued, by his sense of the usefulness, the practical advantages of scientific knowledge. We find an echo of this in the second of these three extracts, "The Hand Dreams," from chapter 4 of *L'Eau et les rêves*, "Water's Compounds." Even when *homo faber* is preoccupied with purely practical ends, his efforts are sustained by his "inward reverie," by his desire to know, this desire being experienced as a true emotion. Underlying these efforts is, it must be noted, the break with immediate reality, with "visual observation," the *"dynamic hand"* being "the antithesis of the *geometric hand.*" Reverie and imagination are, like scientific thought, constituted by a break with the immediate. Indeed, in the penultimate paragraph of *Lautréamont* (1939) Bachelard affirms quite unequivocally that "the new thought and the new poetry require a rupture and a conversion. Life must wish for thought" (155).

Bachelard would seem, at this stage, to be struck by the similarity rather than the difference between the activities of scientists and poets. In *Lautréamont* particularly, he seems anxious to underline this, using mathematical language when discussing poetry—"resultant," "degree of freedom," "group" in this first extract, from chapter 2 "Lautréamont's Bestiary"—comparing mathematics and poetry explicitly when he suggests here the need for a projective poetry parallel to projective geometry, and implicitly, in the idea of dynamic imagination. This is not found in *La Psychanalyse du feu*, where the emphasis is on material imagination, so that one wonders how Bachelard came to formulate it in *Lautréamont*. His reading of *Les Chants de Maldoror*, as the first extract will show, is an excellent example of how he attends sensitively and intelligently to what is in the text. He does not regard it, as many critics have done, as evidence of the poet's insanity; he does not therefore read the text as a sign of something other than itself. Even so, was it simply a close reading of the text that led to Bachelard's discovery of dynamic imagination? Why indeed choose to write about Lautréamont? Curiously, it is the only book he devotes to just one writer, and besides this, it does not fit into his study of the "four poetic elements." Michel Mansuy suggests that the choice of Lautréamont was due to the hazards of book publishing, three editions of Lautréamont's complete works coming out in five months, between April and August 1938 (1967: 56–58). Vincent Therrien dismisses this idea, explaining that Lautréamont was the subject of a philosophy course Bachelard taught at Dijon in 1937 (1970: 46). *Lautréamont* belongs within the context of his work on science, his interest possibly aroused by the mathematical background of this undeniably peculiar and very fashionable poet.[7] If this epistemological context is borne in mind when reading *Lautréamont*, we realize that the previous ten years had accustomed him to the idea of deformation, to the dynamic character of thought, so that Bachelard the reader of poetry was prepared to make a "dynamic interpretation" of Lautréamont's strange images, to accept his "activist," "motory" imagination. Similarly, he was used to thinking in terms of second-order reality, constructed rather than found, and he could as a result read these images not as a mad distortion of immediate reality but as a realizing of the possibilities of reality: a fishtail with wings, a particle behaving like a wave, why not?

After *Lautréamont*, in the four books on the elements, material and dynamic imagination are inseparable. The second of these extracts is a good example of this: the matter the hand dreams is not the matter we use in everyday life; it is matter without form, dreamed in its "active ambivalence," endlessly deformed, rhythmic, dynamic. Similar ideas are found in the third extract, "Isomorphic Images," from chapter 5 of *La Terre et les rêveries du repos* (1948), "The Jonah Complex." The phrase "isomorphic images" again suggests a desire on Bachelard's part to bring science and poetry together, making us think of poetic images in terms of crystal structures and the laws governing them. This notion of isomorphism develops from the idea of a group of metaphors, put forward in a rather abstract way in the extract from *Lautréamont*. Bachelard now explains the coherence of these groups, examined in the three preceding books on water, air, and earth, in terms of the "imagined matter underlying all form." Matter is not just dynamic, it determines form: this is fundamental to his view of poetic imagination. Again, the influence of his work on science can be discerned, enabling him to respond sympathetically to these strange poetic images and take them seriously, to interpret these "compound," even "supracompound" images as going *beyond* perception, *beyond* the phenomena of immediate experience.

Poet and scientist both transcend the familiar, everyday world and its objects, both create what in *La Philosophie du non* (1940) Bachelard calls "*superobjects*" (139). Just as in science the subject is created as project, so in poetry he is, as Bachelard puts it in "The Hand Dreams," "an anagenetic duration"; his "becoming" is a function of "material becoming." Yes of course, poet and scientist approach matter very differently—compare for instance "The Hand Dreams" with Bachelard's description in "Toward a Non-Cartesian Epistemology" of how a modern scientist would study Descartes's drop of wax. Science and poetry are undeniably distinct, yet reading Bachelard we discover how much they share. Both break with immediate experience, both are *against* life, both explore possibility in object and subject, in the world and in human beings.

Extract I
Mathematics and Poetry: On Lautréamont's Dynamic Imagination

In Lautréamont's poetry, the bird is the symbol of all joyful, effortless activity, for it moves so gracefully, so freely. Thus, there are singing birds in these poems, strange though this may seem . . .

There is, moreover, a wide variety of birds here, yet no one bird is singled out as having a special value, no one bird is violently dynamized, with the exception, of course, of the eagle, whose importance surely stems from the close relationship between talon and claw, the eagle being, in fact, a flying claw. It would seem that the air is the region of effortless metamorphosis, where metamorphosis encounters no obstacle. Thus, as if it were perfectly natural, when Maldoror has to go into hiding:

> by the help of a metamorphosis, and without abandoning his burden, he mixes with the flight of other birds.[1]

As the bird flies, far up into the sky, it loses its individuality; it becomes flight, flight in itself.[2] Now, activist imagination makes use of the bird for only one reason, and that is to bring into being free flight, flight having here the sense of escape, of fleeing. Flight, or rather fleeing, belongs to a psychology of a rudimentary kind, and it is therefore given a concrete form by means of schematic metamorphosis.

Here again there is as it were a kind of contamination between birds and fishes, the nature of which will be plain once we apply the method of dynamic interpretation that we are suggesting for Lautréamontism. What we have here is, in fact, an almost geometrical combination of flying and

---

1. Bachelard is quoting from the sixth canto of *Les Chants de Maldoror*. The translation given here is from *The Lay of Maldoror*, translated by John Rodker, privately printed for subscribers only by the Casanova Society (London, 1924), 308.

2. Bachelard's footnote: cf. Paul Eluard, *Donner à voir* (Paris: Gallimard, 1939), 97: "There is no distance, as the bird flies, between clouds and men." André Breton, *Poisson soluble*, published with *Manifestes du Surréalisme* (Paris: Éds. du Sagittaire, 1924), 89: "Birds lose their form after losing their colors. They are reduced to a gossamer existence." Note amended. My translations; these works have not been translated.

swimming. No longer shall we be astounded, no longer shall we think it baroque that the *concrete* resultant of flying and swimming obtained by Lautréamont's essentially realizing imagination should be purely and simply a fishtail with wings, the synthesis of means of propulsion. Nature has taken this kind of realization to its limits and produced the flying fish; Lautréamont's imagination can only produce a flying tail. However crude and puerile this realization may be, it is, we believe, quite sufficient for us to see that Lautréamont's imagination is *natural*. Conversely, the flying fish is one of nature's nightmares.

When the poet has taken it upon himself to schematize realizations in this way, then the power of metamorphosis is at its height. Parts of different beings are put together, as in a nightmare:

> He drew the fish's tail from the well and promised to rejoin it to its lost body if it would announce to its Creator the impotence of his messenger in dominating the furious waves of the maldorian seas. He lent it two albatrosses' wings and the fish tail took to flight.[3]

This fragmented, incongruous, bewildered genesis, resting upon biological chaos, has quite naturally led people to diagnose insanity here or else to accuse the writer of sheer gruesome contrivance. We ought, however, to see in it simply and solely a kind of derangement of the animalizing faculty, which on this occasion animalizes anything and everything. Yet because of its very deficiencies, this immediate biological synthesis shows us very clearly the *need to animalize* from which imagination originally develops. Imagination's first function is to produce animal forms.

Furthermore, we should be far less surprised by what naïve imagination can construct were we to go deep down into dreams, to the very origin of psychic forces, and not look for quick ways of translating the symbols we find in dreams into something human and accessible, as classical psychoanalysis has so often done. Rolland de Renéville has observed, following here the psychologist Chamaussel, that children sometimes confuse birds and fishes.[4] This *confusion*, this fusion, is nonsensical only for a mind that is wholly convinced of the permanence of all forms. A very different view

---

3. *The Lay of Maldoror*, 314.
4. Bachelard is probably referring to Rolland de Renéville's *L'Expérience poétique*. See note 5.

is held by those who see kinetism as the fundamental poetic requirement: between swimming and flying there is an obvious mechanical similarity. Birds and fishes live in something that has volume, whereas we live only on a surface. To use a mathematical expression, they have one degree of freedom more than we have. Since birds and fishes inhabit a similar dynamic space, the confusion of these two kinds of creature is not at all absurd in the realm of psychic forces and of motory imagination. If it is true that poetry comes to life at the sources of language, and that it is one with a fundamental psychic excitation, then basic movements such as swimming, flying, walking, and leaping ought to summon up special kinds of poetry . . .

We can also explain the confusion of birds and fishes in another way. Rolland de Renéville, that dowser of poetic experience, has observed that

> . . . some occultists classify birds and fishes as belonging to a separate race from that to which they assign all other animals. The so-called primitive painters, too, have left us many a landscape in which the creatures that dwell among the leaves of the trees are fish. Last, but most important of all, we can never forget that this strange confusion is evident in the first few lines of the Bible, where we read that God created both birds and fishes on the very same day.[5]

Lautréamont has penetrated deep into the mysteries of the biological dream, unwittingly and as if guided by the light of nature. Lautréamont is indeed a *primitive* in dynamic poetry.

This concept of the *primitive in poetry* is not an easy one and would surely require years of study, by psychologists rather than by experts in literature. We should be much mistaken if we were to look for the elements of this primitivity in the poetry of troubadour and trouvère. If we look at the problem from the standpoint of psychology, we shall soon see that, paradoxically, *primitivity in poetry develops very late*. This probably stems from the fact that, in the realm of language more than anywhere else, intellectual values, objective values, and indeed taught values are quick to oppress. *Primitive poetry* must create its language, it must always be accompanied by the creation of a language, and thus it may well be hampered by the language that has already been learned. Poetic reverie itself will soon

---

5. Bachelard's footnote: Rolland de Renéville, *L'Expérience poétique* (Paris: Gallimard, 1938) 150. Note amended. My translation; this book has not been translated.

turn into scholarly reverie, that is to say, into the reverie learned in the schoolroom. We must rid ourselves of books and of teachers if we are to rediscover *poetic primitivity*.

We shall need real courage if we are to found *projective poetry* before there is ever metrical poetry, just as sheer genius was needed to discover, very late in the day, that beneath metrical geometry there lay projective geometry, which is in fact essential and primitive. Poetry and geometry are completely parallel here. The basic theorem of projective geometry is as follows: what elements of a geometric form can, with impunity, be deformed in a projection in such a way that geometric coherence remains? The basic theorem of *projective poetry* is as follows: *what elements of a poetic form can, with impunity, be deformed by a metaphor in such a way that poetic coherence remains?* In other words, *what are the limits of formal causality?*

Once we have thought for a while about the freedom and the limits of metaphor, we realize that certain poetic images are *projected* onto each other, with precision and accuracy, which means, in fact, that in *projective poetry* they are one and the same image. To take an example, in our study of the Psychoanalysis of fire, we became aware that "images" of inner fire, of hidden fire, of fire that smoulders beneath the ashes, of, in short, all unseen fire which, because it is unseen, requires metaphor, are all of them "images" of life. The projective bond is so primitive here that images of life can be easily translated into images of fire and vice versa, and we may be quite sure that everyone will understand.

The deformation of images must therefore indicate, in a strictly mathematical way, the *group* of metaphors. Once we can determine the different *groups* of metaphors in a particular poem, we shall see that sometimes certain metaphors are unsuccessful because they have been introduced without any concern for the cohesion of the group. Any sensitive poetic spirit will, of course, react spontaneously to these mistaken accretions, without feeling the need for the rather pedantic procedure to which we are referring. Nevertheless, any metapoetics will have to embark upon the classification of metaphors, and sooner or later, it will have to adopt the only really essential method of classification, that is to say, the determination of groups.

Lautréamont (1939; Paris: Corti, 1951), 49–55.

Extract II
The Hand Dreams: On Material Imagination

*F*rom the union of water and earth comes a kind of soft paste, which is in fact one of materialism's basic schemata. We have always thought it rather odd that this should not have been studied by philosophers, for in our view it is the schema of a truly inward materialism, in which form is driven out, erased, and dissolved. Here, all the problems of materialism are posed in their most elementary terms, since intuition has lost its preoccupation with form. The problem of form is now secondary. This soft pasty substance will give us our very first experience of matter.

In all this, the action of water is unmistakable. We can start to think about the particular nature of the substance concerned, whether this be earth or flour or plaster, once we have worked or kneaded it for a while, but when we begin, we can only think about water. Water is our first helper here. It is the activity of water that first sets us dreaming as we work and knead. It is no surprise, then, that water should be dreamed in its active ambivalence. There can be no reverie without ambivalence, and no ambivalence without reverie. Our dreams of water are centered by turns upon its power to soften and its power to agglomerate. Water unbinds and water binds.

Water's first action is very clear. In the words of the old chemistry books, water "tempers the other elements." Water destroys dryness, the work of fire, and so water conquers fire; patiently, it takes its revenge on fire; it soothes away fire, allaying all fever. And water is more powerful than the hammer when it eradicates earth and softens substance.

So the working of clay and the kneading of dough continue. Once water has been forced deep into the very substance of the crushed earth, once the flour has drunk the water and the water has eaten the flour, then there begins the experience of "binding," and the long, long dream of "binding."

This power to *bind* substantively, using common inner bonds, is ascribed sometimes to earth and sometimes to water by the worker dreaming his work. Indeed, in many an unconscious, water is loved for its viscosity. As we experience viscosity, we rediscover a great number of organic images, and these will be the ceaseless preoccupation of the worker as patiently he kneads, and molds, and shapes.

It is for this reason that Michelet is, in our eyes, an adept where this

Dynamic and Material Imagination 103

a priori chemistry is concerned, this chemistry whose basis is unconscious reverie. For Michelet,

> sea-water, even the purest, when procured in the open ocean, free from all admixture of earth, is of a whitish color, and somewhat viscous ... Chemical analyses do not explain this characteristic. It originates in an organic substance, which in attaining they destroy, depriving it of its specialty, and reducing it violently to general elements.[1]

Then quite naturally, Michelet's pen finds the word mucus *and so completes this mixed reverie in which we see both viscosity and mucosity:*

> What is this *mucus* of the Sea, this viscosity which water in general presents? Is it not the universal element of life?[2]

Viscosity is also sometimes the mark of oneiric weariness; it prevents the dreamer from making any progress. We then live out our sticky dreams in viscous surroundings. The dream's kaleidoscope is full of round objects, of slow objects. If it were possible to make a systematic study of these soft dreams, we should discover and come to know a mesomorphic imagination, that is to say, an imagination intermediate between formal imagination and material imagination. In a mesomorphic dream, it is only with the very greatest difficulty that objects take on a form and then they lose it, softening and spreading like dough. The soft, sticky object, indolent and sometimes phosphorescent, though never luminous, is, we believe, indicative of the greatest ontological density of oneiric life. These dreams that are centered on soft pasty substances are dreams which, by turns, do battle or suffer defeat, in order to create, to give and then take away form, to knead and to mold. To quote Victor Hugo, "everything loses form, even that which has no form."[3]

The eye itself, pure vision, grows weary of solids. Its great wish is to dream deformation. If when we look upon the world we could do so with all the freedom of our dreams, then everything would be fluid in an intu-

---

1. Jules Michelet, *The Sea*, trans. W. H. Davenport Adams (London: T. Nelson and Sons, 1875), 92.
2. Ibid., 93.
3. Victor Hugo, *Les Travailleurs de la mer*. My translation here. For a published translation, see *Toilers of the Sea*, trans. W. Moy Thomas, (London: J. M. Dent and Sons; New York: Dutton, 1961).

ition that was truly alive. Salvador Dali's "soft watches" flow and drip over the table's edge. They live in a sticky space-time. Like generalized water clocks, they take something that has been directly subjected to the temptations of monstrosity, and make it ooze and flow. We have only to think about Dali's *Conquest of the Irrational* and we shall understand that this pictorial Heraclitism is the fruit of an extraordinarily sincere reverie. Deformations as profound as these must of necessity make substance their source. In Dali's words, the *soft watch* is flesh, it is "brawn."[4] These deformations are often poorly understood because we look at them as if they were something static. Indeed, there are those *stabilized* critics who would not hesitate to regard them as the work of madness. Such critics do not live the deep oneiric power of deformation, nor do they share in that rich viscous imagination which will from time to time bestow the great gift of divine slowness upon our briefest glance.

We could find many traces of the very same dreams in the prescientific mind. Thus, for Fabricius, pure water is already a kind of glue; it contains a substance to which the unconscious has given the task of *realizing* the binding action at work in soft, pasty substances:

> Water has a viscous and sticky matter which means that it will attach itself with ease to wood and iron, and to other rough bodies.[5]

It is not just some unknown man of science like Fabricius whose thinking makes use of materialist intuitions of this kind. We shall find exactly the same theory in Boerhaave's chemistry. Boerhaave writes that:

> ... so likewise stone and brick, when reduced to powder and distilled, in dry vessels, give out an aqueous moisture, which, as a kind of cement, enters their composition, and sticks their particles together.[6]

Water is, in other words, a universal glue.

---

4. For the published translation, see Salvador Dali, *Conquest of the Irrational*, trans. David Gascoyne (New York: J. Levy, 1935). My translation here.
5. Jean-Albert Fabricius, *Théologie de l'eau, ou essai sur la bonté, la sagesse et la puissance de Dieu, manifestée dans la création de l'eau*, translated from the German (Paris, 1743). My translation here; no English translation appears to have been published.
6. Boerhaave, *A New Method of Chemistry, including the History, Theory and Practice of the Art*, translated from the Latin, *Elementiae Chemiae*, by Peter Shaw, M.D., 2d ed. (London: T. Longman, 1741).

Dynamic and Material Imagination

This *adherence* of water to matter cannot be understood if we limit ourselves to visual observation alone. To this we must add observation through touch. This phrase makes reference to two of our senses. It is interesting to study the action of tactile experience when it accompanies visual observation, however minor this action may be. We shall thus be able to rectify the theory of *homo faber*, which is far too quick to assume an exact parallel between worker and geometer, between action and vision.

We suggest, therefore, that both the remotest reverie and the harshest toil be reintegrated into the psychology of *homo faber*. The hand has its dreams, too, and its own hypotheses. It helps us to come to know matter in its secret, inward parts. The hand, then, helps us to dream matter. The hypotheses of "naïve chemistry" which stem from the work of *homo faber* are at least as important psychologically as the ideas of "natural geometry." Indeed, since these hypotheses conceive of matter more intimately, they give greater depth to reverie. When we mold, and knead, and shape, there is no longer any geometry, nor any sharp edge or interruption. It is indeed an endless dream, an activity in which we close our eyes, an inward reverie. And it is, moreover, a rhythmic activity, whose precise, insistent rhythm takes possession of our whole body. A truly vital activity, then, and one which shares the chief characteristic of duration: rhythm.

This reverie that comes into being whenever we handle any kind of soft paste is necessarily in entire agreement with a particular kind of will to power, with the virile pleasure of *penetrating* substance, of *touching the inward parts* of substance, and coming to know what lies within the seed, conquering the earth from within, just as water conquers earth. Here, too, is the joy of rediscovering an elementary force, of sharing in the war between the elements, in the power to dissolve, irremediably. Then the binding action begins, and as we shape, and knead, and model, progressing slowly but steadily, we experience a very special kind of joy, less satanic than the joy of dissolving. Our hands are directly aware that, very gradually, the union of earth and water is taking place. A different kind of duration is now established in matter, a duration in which there is no interruption, no momentum, and no definite end in view. This duration is not therefore *formed*. It lacks the various *stations* of successive attempts at form, which we should find were we to consider work on solids. This duration is a material becoming, a becoming which has a deep and inner source. It offers too an objective example of inward duration. Poor, simple, and crude as this duration is, hard work will be needed if we are ever to reach it. Even so, it is an anagenetic duration, rising upward and producing: a laborious duration indeed. All real workers are those who have "lent a hand." Theirs is an operative will, a manual will. This very special kind of will can be seen in the structure of our hands. Only someone who has crushed grapes and blackcurrants will understand the hymn to Soma: "our ten fingers spur on the charger in the vat." If the Buddha has a hundred arms, it is because he handles and shapes matter.

These soft pasty substances produce the *dynamic hand* which is almost the antithesis of the *geometric hand* of Bergson's *homo faber*. This dynamic hand is no longer an organ of form but of energy. It symbolizes the imagination of force.

Were we to think about all the different crafts or trades in which kneading and modeling play a part, we should gain a much better understanding of the *material cause*, and see it in all its variety. The activity of modeling is not satisfactorily analyzed by the attribution of forms. Nor is the effect of matter satisfactorily indicated by the way it resists our modeling. Any activity which involves handling some kind of soft paste will lead to the idea of a truly positive, truly active material cause. What we have here is a natural *projection*. It is a particular instance of that *projecting* thought which carries all thought, all action, all reverie from man to things, from worker to his work. The theory of the Bergsonian *homo faber* can envisage the *projection* only of clear ideas. It has taken absolutely no notice of the *projection* of dreams. Crafts which carve and cut cannot teach us about matter in its inward, secret parts. Projection must remain external and geometric here. The role of matter cannot even be to support action. Matter is but the residue of acts, that which has not been cut or carved away. The sculptor standing before his piece of marble is the punctilious servant of the formal cause. He finds form by eliminating the formless. The modeler with his lump of clay finds form by deforming it, by the dreamy germination of the amorphous. It is the modeler who is closest to the inward, germinating dream.

L'Eau et les rêves: essai sur l'imagination de la matière (Paris: Corti, 1942), 142–48.

Extract III
Isomorphic Images

All the great images that tell of human depths, the depths of which man is aware in himself, in things, and in the universe, are, as we shall see, isomorphic images. It is for this reason that they so readily serve as natural metaphors of each other. The word *isomorphism* may well seem inappropriate with regard to this kind of correspondence, occurring as it does at a moment when these *isomorphic* images in fact lose their form. Yet this loss of form is still dependent on form, and indeed it explains form. The two very different dreams of sheltering in the oneiric house and of returning to the mother's body reveal, then, the selfsame need for protection. Here, we remember a phrase of Claudel's, hyphening, as it were, these two dreams: "a roof is a belly."[1] Ribemont-Dessaignes says, even more explicitly, in *Ecce Homo*:

> And the room surrounds them like a belly,
> Like the belly of some monster,
> And already the beast digests them,
> Deep in eternal depths.[2]

This isomorphism of lost forms will take on its full meaning if the reader comes with us into our chosen field of study, and considers in a systematic fashion the *imagined matter* underlying all form.[3] The reader will then discover a kind of *materialized repose*, the paradoxical dynamics of warmth that is both gentle and still. There is, it would seem, a *substance of the depths*. And now are we absorbed into the depths . . .

1. Bachelard's footnote: Claudel also says, in *Tête d'Or* (Paris: Librairie de l'art indépendant, 1890): "and I came forth from the belly of the house." Again, he says: "the house commands, like the belly we cannot disobey." My translations. *Tête d'Or* has been translated by John Strong (New Haven: Yale University Press, 1919). However, the translator omits the word "belly" from the first phrase, rendering it as: 'I went out from the house and left the old familiar faces" (9); the second phrase he omits entirely, leaving out several of Claudel's lines (14).
2. Georges Ribemont-Dessaignes, *Ecce Homo* (Paris: Gallimard, 1945). My translation; this poem has not been translated.
3. Bachelard's footnote: in the Vedic hymn to the wood cabin, we find lines in which it is compared to a pot belly.

Let us offer an example of this isomorphism of substance. This substance of the depths is, in point of fact, the *enclosed night* of cavern, belly, and cellar. Joë Bousquet, describing night in a very fine article of his in the review *Labyrinthe*, speaks of materially active night, as penetrating as a corrosive salt. This "salt night" is also the subterranean night *secreted* by the earth, the cavernous night at work deep within living bodies. Thus, Joë Bousquet will evoke "night, alive and voracious, to which everything that has breath is bound, deep within its being." Straightaway, in this very first notation, we feel we are now far beyond the familiar realm of images formed in perception. It is to material imagination that we must turn if we would thus transcend the night, and reach the world beyond the phenomenon of night. Then might we lift the dark veil of night, and see, as Bousquet puts it, the night beyond darkness:

> Other men can only think of it with fear, they have no words with which to speak of it. It cannot be dispersed, and like a fist, it closes upon all that issues forth from space. It is night before flesh ever was, and it looks at man with flowering eyes, whose mineral and bewitching color is rooted in that same darkness to which belong plants, and flowing hair, and all the seas.[4]

Before flesh, and yet within flesh, indeed in that fleshly limbo where death is resurrection, and where eyes flower afresh, in their astonishment...

We have said several times that images which second-rate poetry would refuse to associate will, in such a web of interwoven images as this, merge into one another, by virtue of some kind of oneiric communion. Here flowing hair will know the night of caves that lie beneath the sea, just as the sea knows the plant's subterranean dreams. The dark night of the depths summons all these images not to the firmament's vast, dark unity, but rather to that matter made of darkness itself, which is the earth, earth which the very roots digest. Whether we digest or bury in the earth, we follow in the path of the very same transcendence; here, Jean Wahl is our guide, though we are probably interpreting his words more materially than he would wish:

> In the depths where a man can be so very comfortable,
> In the primal clay of the flesh ...

---

4. My translation.

> I am plunged deep...
> In the unknown land, where my unknowing brings the dawn.[5]

Joë Bousquet's article expresses, in many different ways, this fleshly prison of the night, in relation to which Jonah is but a tale that is too naïvely told. Speaking of the poet, Joë Bousquet says:

> His body, like our own, envelops an active night, which swallows up all that is yet to be born, and even so, this sulphuric night he allows to devour him too.

If you are prepared to linger and dwell in all these images, and then let each slip slowly into the other, you will come to know the extraordinary delights of these compound images, images that at one and the same time serve differing demands of the life of our imagination. This is in fact the peculiar characteristic of the *new literary mind,* so typical of contemporary literature, in that it changes its *level of imagery,* rising or falling along an axis which runs, in both directions, from the organic to the mental and spiritual, and is never content with just one plane of reality. Thus, the *literary image* is privileged in that it acts as both an image and an idea. It can imply something intimate and something objective. And so it is hardly surprising to find that the literary image is at the heart of the problem of language.

We can now understand how Joë Bousquet can say that "his flesh's innermost shadow binds [the poet] under the spell of what he sees" or, even more swiftly, that the poet "binds himself under the spell of things." By using the reflexive, Joë Bousquet gives a new meaning to the idea of being spellbound, although the reflexive verb *to bind oneself under the spell of* still points toward the external world; it bears, then, the double mark of introversion and extraversion. "To bind oneself under the spell of" is therefore one of those very rare expressions which combine the two fundamental movements of the imagination. The most external of images—day and night—can thus become images of the innermost world. It is here, in this world deep within us, that these great images find their powers of conviction. Externally, they would continue to be a means of explicit communication between minds. Communication through the inner world

---

5. Bachelard gives his source as Jean Wahl, *Poèmes* (Montréal: Éds. de l'Arbre, 1945), 33. My translation; these poems have not been translated.

has, however, a much higher value. *Jonah*, the oneiric house, and the cave that we imagine are all archetypes which do not need to be experienced in real life in order to have their effect on every single soul. Night binds us under its spell, and the darkness of cave and cellar takes possession of us, like the womb. We have only to touch upon just one small facet of these compound or supracompound images, whose roots plunge deep into man's unconscious, and immediately, the very slightest tremor will reverberate and be everywhere reechoed.

La Terre et les rêveries du repos: essai sur les images de l'intimité (Paris: Corti, 1948), 173–77.

In the realm of language more than anywhere else, intellectual values, objective values, and indeed taught values are quick to oppress.

It is the activity of water that first sets us dreaming as we work and knead . . . Our dreams of water are centered by turns upon its power to soften and its power to agglomerate.

The peculiar characteristic of the *new literary mind* . . . [is] that it changes its *level of imagery,* rising or falling along an axis which runs, in both directions, from the organic to the mental and spiritual, and is never content with just one plane of reality.

These three statements, from each of the three extracts here, take us to the heart of Bachelard's conception of poetic imagination. He values imagination as liberating the subject, but this liberation is not at the subject's behest. It is matter, its possibilities and "powers," that frees him. The *"new literary mind,"* like the new scientific mind—the phrase is surely used in order that this comparison be made—has broken with the immediate, the simple, the identical, and into difference. "Imagination can overleap extraordinary differences," Bachelard says (*TRV* 291), and this perhaps explains his preoccupation with poetry. "A poem," he writes, "is essentially *an aspiration to new images*. It corresponds to the essential need for *newness* that characterizes the human psyche" (*AS* 8). Poetic imagination is not for Bachelard frivolous escapism, from which we must come smartly back to reality. Submission to immediate reality mutilates the human being; modern science, with its "excellent mental hygiene," restores us to health (*NES* 88), but poetry remains a more accessible way of attaining full humanity. To quote *L'Eau et les rêves*: "Imagination . . . is a faculty of superhumanity. Man is man in proportion as he is superman. Man should be defined by the group of tendencies that drive him to go beyond the *human condition*" (23).

Chapter 7

Reading Poetry: 1938–1948

> It would seem that the reader is called upon to *continue* the writer's images; he is aware of being in a state of open imagination.
> —*La Terre et les rêveries du repos*

Bachelard often described himself as "a reader, simply and solely a reader" (*TRV* 6), not out of modesty alone but because he was convinced of the value of reading. He did not regard himself as a literary critic or even a literary theorist, and in his opinion the activities of reading and criticism remain distinct. Yet he was not simply a reader; he was also a writer. In his books on poetry, he writes what he reads, for us and for himself. This last point is important. Bachelard describes himself reading "pen in hand" (*AS* 283) and enjoins us to do the same, either copying out the poem we read—far better he says than reading it aloud—or in our own way "continuing" the writer's images. Reading and writing are for him inseparable, but why this should be is not immediately evident. The four extracts in this chapter have therefore been selected in view not so much of his theory of imagination as of his practice of poetry. They will, in some measure, help us to understand both how and why he reads.

Certain kinds of reading earn Bachelard's disapproval. Literary critics read "pen in hand," yet he is intolerant of literary criticism. He is at his most vehement in *Lautréamont* (1939) and in an article of 1942 on Jean Paulhan's *Les Fleurs de Tarbes* (*DR* 176–85), his target being those dogmatic critics who approach a text with certain preconceived ideas, in terms of which they interpret and evaluate it, and who consequently fail to read the text itself. The dogmatic critics he has in mind belong, as these extracts suggest, to two categories, to that of traditional, academic criticism in France, and also to the more recent psychoanalytical approach.

In the former, literature is assumed to be the mirror of rational thought and conscious life. For example, Shelley's *Prometheus* (Extract II) is usually considered from this point of view, as expressing social and political ideas. Readers tend, therefore, to approach this poem with their minds already made up, knowing what it is about, reading what they want to read rather than the poem itself, so that they miss the aerial, cosmic dimension that Bachelard reads here, and the poem's *human* value. Similarly, from this point of view, Orlando's tree (Extract III) would be automatically interpreted as just a tree in a remembered landscape from the author's life, and readers would pass on, unaware of "this communion of hard objects encircling a core of hardness," closing themselves to both the power of Virginia Woolf's "material imagination" and the benefit Bachelard believes it has for everyone. When, for instance, it encounters the Narcissus theme (Extract I), or these images of the smithy (Extract IV), this traditional literary criticism may well be nonplussed. Here, psychoanalytical criticism comes into its own, with its particular preconception of literature as the mirror of unconscious life and the irrational.

Bachelard is especially wary of the stock responses of psychoanalytical criticism. It has much to offer in its understanding of poetic images—he describes in *L'Eau et les rêves* how it helped him to read nonpositively (80–81)—yet its interpretations are, in his view, both limited and automatic, so that too often this kind of reading is simply a conditioned reflex, with the reader reducing the text to fit a preexisting framework of ideas. In this perspective, Narcissus is and will always be a passive, negative, neurotic figure: whenever this name occurs in literature, we know how to respond, and we need not read on. Again, if we adopt this approach, we shall miss both the essential, inherent complexity of Narcissus which Bachelard discerns, and also the variations on this theme, the different ways in which this figure is used by different writers. Here too, we fail to read the text itself. As for images of the smithy and of tempering, a psychoanalytical critic would doubtless interpret them in sexual terms, his reflexes springing into action when he sniffs sadism, preventing him, once again, from reading what Bachelard suggests is there to be read, the interplay of the material and the moral, the cosmic and the human.

Literary critics are severely censured for their failure to read the text itself. We therefore expect Bachelard's approach to be that of "close reading," of submission to the text. However, as these four passages show, this is seldom what he does. He does not restrict himself to Valéry's Narcissus, for example, or to Lönnrot's smith; he links them to similar themes and images in other works, turning a blind eye to the way a writer develops his theme in one particular work, and to the pattern of different images within one text. When he does limit himself to one text or to one writer, Virginia Woolf's *Orlando*, for instance, or Shelley's *Prometheus*, we note two things: first, that he reads only fragments, not the whole text, and second, that he tends to move away from the text, exploring not the text but his own response to it. There would appear to be a discrepancy between Bachelard's intention as a reader and his achievement. He advocates slow, careful reading and rereading *"line by line"* (*TRV* 6), yet in practice, his own reading is not subordinate to the text but very free. While this indiscipline on Bachelard's part has vexed and provoked many, it is, paradoxically perhaps, central to his understanding of the activity of reading. Charles Mauron (1962: 29) and Jean-Pierre Roy (1977: 8, 208, 212, 217), different though their own criteria may be, both attack him for being "unscientific." Jean Ricardou (1968: 214, 221) is more damning when he describes Bachelard's reading as "frivolous," "sentimental mush," and Mary Ann Caws (1966: 28), who is otherwise sympathetic to his approach, is critical of his tendency to "ramble." What all of them forget is that Bachelard did not set out to be either a literary critic or an "objective reader": he is someone who reads "pen in hand," a reader who writes. The distinction is crucial.

To understand it, we must remember Bachelard's attitude to literary criticism. What disturbed him was not so much that the critic judges a work, but that he reduces the work to the terms of his judgment, that is to say the work has to fit into a preexisting framework of ideas; it is perceived through the grid of what is already known, so that anything new or different is suppressed. Chief among the critic's preconceptions is that the work mirrors the life, that he must therefore in judging it reduce the work to the life. For Bachelard, this is illogical; it forgets the simple fact that poems—and indeed all literature, all works of art—have been created, that they imply a fundamental break with life: "the work of genius," he declares in *Lautréamont*, "is the antithesis of life" (102). Critics also forget another simple fact, namely that literature is written, that the images of poetry are not the images we perceive in our daily lives, nor are they the images of our dreams and our unconscious: they are, first and foremost, written images. This is a constant theme in Bachelard's books on poetry, and it explains why he is not just a reader but a reader who writes.

How Bachelard reads and the value he places on reading are determined by his awareness—an awareness that grows markedly more sharp over the years—that poetry is language, and more especially new language. "The function of literature," he will declare in *La Terre et les rêveries de la volonté* at the end of the decade, "and the function of poetry is to give new life to language by creating new images ... Already there seem to be areas where literature is an *explosion of language*" (6–7). This takes up something he said a few years earlier, in *L'Air et les songes*: "a literary image is an explosive. It will suddenly shatter ready-made phrases" (285). The writer breaks with everything that already exists, with his inner life just as much as with his ideas (288). Creation, for Bachelard, always signals a break; it is not repetition but "rupture" (*L* 155). These explosions and rendings occur because language is polyphonic and polysemic (*AS* 283), because, as he puts it, there is in every word "the desire for alterity, for double meaning, for metaphor" (*AS* 10). Bachelard does not pursue the mechanics of polyphony and polysemy from a theoretical point of view. He is not, and has no ambition to be, an expert in linguistics. What concerns him is polyphony at work in the text, and more than this, the practice of polyphony and polysemy not only by the writer but the reader.

There are a number of examples in the four extracts chosen in this chapter of the ease with which words break from familiar referents and couple with strangers: Mallarmé's "severe fount," his "remembrances which are like leaves beneath your ice's profoundness," Valéry's "Rose of the Water," Shelley's "sandals of lightning" and "dizzy arches," Virginia Woolf's description of Orlando's "floating heart." Bachelard's interest here lies not so much in the ability of language to make sense, however outlandish its juxtapositions, as in its power to free us from conventional, habitual usage, from the "paralysis" of received meanings which deny the true nature of language, its polysemy, and conceal from us its sonority, its polyphony. If we learn to read with what Bachelard calls "a *consciousness of language*" (*TRV* 7), we shall discover its possibilities, its natural creativity. And more important, from this "*consciousness of language,*" he says, "we receive a new psychic dynamism" (*TRV* 7), for in breaking with our own everyday linguistic habits, we ourselves are changed, made different by our reading.

Reading, for Bachelard, is valued not for its revelations about others—how they think, or feel, or write—but for the possibilities it releases in the reader. This attitude in fact predates Bachelard's work on poetry, and is perhaps most evident in his approach to Roupnel's *Siloë*, an approach he obviously felt he had to justify in his introduction to *L'Intuition de l'instant*. His aim, he explains, has not been to "summarize" but rather to "develop" Roupnel's book, to "experience" its intuitions for himself, to learn a personal lesson, with the result that *L'Intuition de l'instant* is not an "objective exposé" but his "experience" of Roupnel's book (7–8). This desire to "develop" and "experience" what others have written underlies Bachelard's insistence on reading "pen in hand," on writing what he reads, so allowing the poet's language to pursue its sonorities through his own language. Close reading, obedience to the discipline of the text, deprives us, he would say, of the real benefit of literature. It imprisons us in something that already exists, it fixes us, and reduces us to identity. Just as the writer breaks with everyday life and language, so the reader must break with the text, not of course ignoring the text—that would be nonsense—but exploring his own response to it by writing. Hence Bachelard's approach to reading, hence too his style. Like the poet, he exploits the possibilities of language: he loves to bring together words from alien contexts, with poetry and mathematics his favorites, Shelley's images, for instance, being "elevation operators." He juxtaposes nouns and adjectives that habit keeps apart, in "floral dynamics," for example; he is fond of learned or technical words, sending his reader back to the dictionary with a term like "intussusception." We must work at Bachelard's language, and in doing so, we gain surprising insights. This use of "intussusception" makes poetry a process both natural and unnatural, gentle and aggressive, while "floral dynamics" provokes an endless collision and inversion of noun and adjective, energy and color, movement and perfume, growth, matter, petals, atoms . . .

Bachelard's sensitivity to language, as a reader and as a writer, is well illustrated by these extracts. His "consciousness of language" is not, though, an experience of decentering; it does not lead to a denial of the reading subject. On the contrary, Bachelard always insists that "the poet must create his reader" (*L* 103), that "the chief function of poetry is to transform us" (*L* 105). This creation, this transformation, is the work of the language we read, of image and metaphor, of written language which is ambivalent, polysemic, polyphonous, always "open," always "new." Reading "pen in hand" means that we are doubly exposed to language, to language, first of all, that is not our own, that is different, and then to language which is at one and the same time ours and not ours, language which did not exist before our encounter with this new linguistic object, yet which does not simply repeat that object but in its own way "continues" it. Writing what we read helps us to understand ourselves, to grasp something of our own complexity, playing as it does between the poles of subject and object, so breaching the dikes of the internal and external worlds.

Extract I
Narcissus

"*Images*" *whose substance or whose pretext is water do not have the permanence and solidity of images provided by the earth, by crystals, metals, and gems. Nor do they have the vigorous life of images of fire. The waters do not fabricate "true lies." Only a really disturbed spirit will be misled by the mirages of river and stream. These gentle water wraiths are usually the work of mere illusion, of an imagination which finds them entertaining, an imagination seeking entertainment. The phenomena of water dancing in spring sunshine offer us an abundance of the common, easy metaphors that pervade second-rate poetry. Inferior poets overwork images like these. It would not be at all difficult to make a collection of lines of poetry in which young water sprites play endless games with very old images.*

Images such as these, however natural they may be, do not enthrall us. They do not arouse deep emotions, as do certain images of fire and earth, even though these are every bit as common. Fleeting images, they leave but a fleeting impression. One glance at the sunlit sky will restore us to all the certainties of light; a decision reached within us or some sudden resolution will restore us to the earth with all its decisiveness, and to the very positive task of digging and building. Almost automatically, by virtue of the fate that rules brute matter, earthly life will reconquer the dreamer for whom glinting water is no more than an excuse for holidays and dreams. The material imagination of water is always in danger, always likely to disappear when the material imagination of earth or fire comes upon the scene. The psychoanalysis of water images is therefore seldom necessary, since these images will vanish, as if spontaneously. It is not every dreamer who is sensitive to their charm. Yet, as we shall see in other chapters, there are forms born of water that are more attractive, more insistent, more consistent: here, deeper and more material reverie is intervening, our innermost being is more deeply involved and our imagination's dreams are closer now to acts of creation. Suddenly, there appears that poetic force which was absent from the poetry of dancing water; the water grows heavy, and dark, and deep, and is materialized. It is in this way that materializing reverie links dreams of water to less fleeting and much more sensual reverie, and in the end constructs upon water and feels water more intensely and more profoundly.

Yet we should have but a poor measure of the "materiality" of particular water images and of the "density" of particular water spirits had we not first studied those iridescent forms that belong only to surfaces. This *density* distinguishing superficial from profound poetry can be experienced when we pass from values to which our response is *sensory* to values to which our response is *sensual*. It is our belief that the theory of imagination can be clarified only if we make a proper classification of sensual values in relation to those that are sensory. Only sensual values will produce "correspondences," for the latter can only translate.[1] It is because we have confused the sensory and the sensual, and assumed there to be a correspondence between *sensations* (these being, in fact, highly intellectual), that we have lost the chance of ever making a truly dynamic study of poetic emotion. Let us begin, then, with that least sensual of sensations, vision, and see how it is sensualized. Let us begin our study of water by considering it in its *adornment*, in the way it presents itself to our gaze. Then very gradually we shall discover, if we are attentive to minute detail, that there is in water *the will to appear*, or at least that it symbolizes the *will to appear* of the dreamer who gazes upon it. We do not consider that psychoanalysis has, where narcissism is concerned, placed equal stress on both terms of the dialectic *to see and to show oneself*. The poetics of water will allow us to contribute to this twofold study.

It was through real insight into the psychological role of natural experience, and not because of a fondness for rather obvious mythology, that psychoanalysis came to mark with the sign of Narcissus the love of a man for his own image, for the face reflected in still waters. Indeed, the human face is first and foremost an instrument of seduction. As he looks at his reflection, man prepares, sharpens, and burnishes face, eyes, and all his tools

---

1. Bachelard's use of "correspondences" is a reference to Baudelaire's poem "Correspondances" in *Les Fleurs du mal* (1857). This poem is often described as initiating symbolism, according to which everything on earth symbolizes some spiritual reality, all things being related because they represent the spiritual essence of the universe. Bachelard's reference does not mean that he is a symbolist—there is no evidence for this—but instead allows him to exploit the idea of relatedness, of patterns of poetic images, and to raise questions about the source of these patterns.

of seduction. The mirror is the *Kriegspiel* of love that is on the offensive. We would draw attention here, though very briefly, to that *active narcissism* forgotten by classical psychoanalysis. We should need a whole book if we were to develop the "psychology of the mirror." We shall have to content ourselves, as we begin this study, with pointing up the profound ambivalence of narcissism, passing as it does from masochism to sadism, living out a contemplation full of regret and then another, this time full of hope, a contemplation that consoles, and another that attacks. To the being who stands before the mirror, we can always put the twofold question: for whom are you reflected? against whom are you reflected? Are you aware here of your beauty, or of your strength? These brief remarks will suffice to show that narcissism is initially complex in character. We shall see in the course of this chapter that from one page to the next, narcissism grows more complex.

We must understand, first of all, that the mirror of water is psychologically useful: water helps to *naturalize* our image, and restore a little innocence and naturalness to the pride of our inward contemplation. The mirror is too civilized an object, too easily handled and too geometric; it is too obviously the tool of our dreaming for it to adapt quite spontaneously to oneiric life. Louis Lavelle has observed the natural depth of reflections in water, and the infinite dreams they suggest, as we see in this quotation from the delightfully poetic preface to a book we find deeply and morally moving:

> If we imagine Narcissus as standing in front of a mirror, then the resistance of mirror and metal will bar his way in whatever he tries to do. He bangs his head against this barrier, and pounds it with his fists; if he walks right round it, he will find nothing at all. The mirror is the prison of a distant world that eludes him, a world in which he can see himself without being able to grasp hold of himself, a world which is separated from him by a false distance that he can diminish but not overcome. The spring is, however, the very opposite, for it is a path that lies open before him . . .[2]

The spring's mirror offers, then, an opportunity for *open imagination*. Its rather imprecise and pale reflections suggest something idealized. As Narcissus gazes into the water reflecting his image, he feels that his beauty

---

2. Bachelard's footnote: Louis Lavelle, *L'Erreur de Narcisse* (Paris: Grasset, 1939), 11. Note amended. My translation; this work has not been translated.

is *continued*, that it is not yet complete, that it must indeed be continued. Mirrors made of glass will, in the bright light of the bedchamber, offer too stable an image. They will be restored to life and naturalness once we can make a comparison between them and water that is natural and alive, when *renaturalized* imagination can admit the *participation* of all the spectacles of river and spring.

What we see here is an aspect of the *natural dream*, for the dream feels the need to etch itself deeply into nature. We cannot dream deeply with *objects*. If we are to dream deeply, then we must dream with *matter*. A poet who starts with a *mirror* must in due course come to the *waters of a little spring*, if he wishes to express his *complete poetic experience*. Poetic experience must, in our view, be placed under the rule of oneiric experience. Poetry as finely wrought as that of Mallarmé is very seldom in breach of this law; it shows us the intussusception of water images in the images of the mirror:

> Oh mirror!
> A cold water frozen with ennui in your frame,
> How often, for how long, unvisited
> Of dreams, and seeking my remembrances which are
> Like leaves beneath your ice's profoundness
> I to myself appeared a far-off shade.
> But ah! Some evenings in your severe fount
> I of my sparse dreams have known the nudity![3]

A systematic study of *mirrors* in the work of Georges Rodenbach would lead to this same conclusion.[4] If we leave aside the window-mirror,[5] that ever clear, ever offensive inquisitorial eye, we see that all Rodenbach's mirrors are veiled, that they all have the same gray life as the waters of the canals encircling Bruges. In Bruges, every mirror is a stretch of still water.

---

3. Stéphane Mallarmé, *Herodias*, trans. Roger Fry (London and Binghampton: Vision Press, 1952).

4. Georges Rodenbach (1855–98), the Belgian symbolist writer, is best known for his novel *Bruges-la-Morte* (1892). For an excellent introduction to Rodenbach, see Philip Mosley, "The Soul's Interior Spectacle: Rodenbach and *Bruges-la-Morte*," *Strathclyde Modern Language Studies* 9 (1989): 25–40.

5. This is an adjustable mirror fixed outside a window, allowing someone indoors to observe the street.

Narcissus goes, then, to the secret spring, deep within the woods. Only there can he feel that he is *naturally* doubled; he opens his arms, plunging his hands into his own image, and speaks to his own voice. Echo is not a nymph who dwells afar. She is within the spring. She is always with Narcissus. She is Narcissus. She has his voice. She has his face. He does not hear her in a great shout. He first hears her in a faint murmuring, the murmuring of his own seductive voice, his seducer's voice. As he gazes upon the waters, Narcissus discovers his identity and his duality, he discovers his dual powers of masculinity and femininity, and above all, his reality and his ideality.

Thus, an *idealizing narcissism* is born at the spring's edge, and we should like to suggest, though briefly, its importance with regard to a psychology of the imagination. We consider this to be all the more necessary since classical psychoanalysis appears to underestimate the part played by this idealization. Narcissism is not, in fact, always a source of neurosis. It has also a positive part to play in art and, by means of rapid transpositions, in literature. Sublimation is not always the negation of a desire; it is not always to be seen as sublimation *against* instincts. It can well be sublimation *for* an ideal. Here, Narcissus will cease to say "I love myself as I am" and say instead "I am as I love myself." I am effervescently because I love myself fervently. I wish to appear, therefore I must adorn my appearance. Thus, life is made rich and strange, and overlaid with a thousand images. Life grows; it transforms being; life takes on a new whiteness; it bursts into bloom; imagination is open to the most remote of metaphors; it shares in the life of every flower. With this floral dynamics, real life gains new strength and vigor. Real life will be all the healthier if it is allowed to holiday in unreality.

This idealizing narcissism expresses, then, the sublimation of the caress. The image we gaze upon in the waters is the contour of a wholly visual caress. It does not need a caressing hand. Narcissus delights in a caress that is linear, latent, and formalized. Nothing material remains in this delicate and fragile image. Narcissus holds his breath:

> My least of sighs
> Would ravish away
> All that I prize
> Upon the pale water

The forests and skies
And Rose of the Water.⁶

L'Eau et les rêves: essai sur l'imagination de la matière (Paris: Corti, 1942), 29–35.

Extract II
Shelley's Prometheus: Images of Air

*S*helley *must have loved all nature, and in his celebration of river and sea he surpasses other poets. His tragic life has bound him for ever to the destiny of water. For us, however, it is the* air *that marks his work most profoundly, and if we were allowed only one adjective with which to describe a poet's work, then there would probably be little disagreement that Shelley's poetry is* aerial. *Yet however apt this adjective may be, it remains inadequate. We wish to prove that Shelley is, materially and dynamically, a poet of the aerial substance. The beings of the air, such as wind, perfume, and light, beings that have no form, all act* directly *on Shelley:*

> The wind, the light, the air, the smell of a flower affects me with violent emotions.¹

If we reflect upon Shelley's work, we shall see that, in some men, the soul responds to the violence of gentleness, *sensitive as it is to the weight of all things weightless; we understand too how souls like this are dynamized by their very sublimation.*

Shelley's poetic reverie bears the stamp of that oneiric sincerity *which is, we would argue, decisive in poetry, and we shall in due course offer*

6. Paul Valéry, *The Narcissus Cantata. Libretto*, Scene 2; see Vol. 3 of *The Collected Works of Paul Valéry, Plays*, trans. David Paul and Robert Fitzgerald (London: Routledge and Kegan Paul, 1960), 325.
1. Bachelard's footnote attributes this quotation to Louis Cazamian, *Études de psychologie littéraire*. Cazamian is in fact quoting—in French—from a letter from Shelley to Claire Clairmont, otherwise Clara Mary Jane Clairmont, written at Pisa on 16 January 1821. I quote here from this letter, published in *The Letters of Percy Bysshe Shelley*, ed. Roger Ingpen (London: G. Bell and Sons, 1914), 843.

both direct and indirect proof of this. Let us now establish the terms of our discussion and take an image in which the "oneiric wing" is obvious:

> Whence come you so wild and so fleet,
> For sandals of lightning are on your feet,
> And your wings are soft and swift as thought.[2]

The images here shift very slightly, detaching the wings from the sandals of lightning, yet this shift cannot destroy the unity of the image; the image is indeed one and indivisible, and what is soft and swift is the movement, and not the wing or the wing's feathers that some dreamer's hand caresses. Once again, we must insist that an image like this does not admit of any allegorical interpretation; it is as a movement of the imagination that our enraptured soul must understand it. Indeed, we would go as far as to say that this image is an action of the soul, that we shall understand if we are willing to *undertake* it. Elsewhere, we read that

> an antelope
> In the suspended impulse of its lightness,
> Were less aethereally light.[3]

Shelley presents us with a hieroglyph that the imagination of forms would have great difficulty in deciphering. It is dynamic imagination that provides the key here: "suspended impulse" is, in point of fact, oneiric flight. Only a poet can explain another poet. Alongside this "suspended impulse" that leaves the imprint of its flight in us, we might place these three lines from Rilke:

> There, where no path was ever made
> We flew.
> The arc is still imprinted on our minds.[4]

---

2. *Prometheus Unbound* 4. 89–91. Bachelard quotes Shelley in the French translation by Rabbe, without referring to the specific poem.
3. *Epipsychidion*, lines 75–77. Again, Bachelard does not refer to this poem by name, giving instead the appropriate page reference in Rabbe's translation.
4. Bachelard's footnote: Rilke, *Poèmes pour Lou*, translated into French by Lou Albert-Lasard (Paris: Gallimard, 1938), poem 6. Note amended. My translation, since these poems do not appear to have been published in English.

Now that we have seen something of what is fundamental and distinctive in Shelley, let us make a more detailed study of the deepest sources of his poems. Take for instance his *Prometheus Unbound*. We shall very soon see that this is an *aerial Prometheus*. If the Titan is bound to the mountain-top, it is so that he may receive the very life of the air. He strains toward the heights with *all the strength* of his chains. He possesses to perfection the dynamics of his *aspirations*.

It is very probable that when Shelley, with all his humanitarian aspirations, dreamed his clear dreams of a happier human race, he saw in Prometheus a being who *raises* man to stand and confront Destiny, and the very gods themselves. All Shelley's demands for social justice are present and active in his work. Nevertheless, the imagination—whether we are talking about its resources or its movements—is always completely independent of any social commitment. Indeed, we are convinced that the real poetic force of *Prometheus Unbound* has absolutely nothing to do with any kind of social symbolism. The imagination is, in some spirits, more cosmic than social. This in our opinion is true of Shelley. Gods and demigods are not so much people—clear images, more or less, of human beings—as *psychic forces* which will play their part in a Cosmos possessing a truly psychic destiny. Let no one hasten to say at this point that these characters must therefore be *abstractions:* instead, this *force of psychic elevation,* this supreme example of the Promethean force, is preeminently concrete. It corresponds to a psychic *operation* with which Shelley was very familiar, and which he wanted to convey to his reader.

Let us remember first of all that *Prometheus Unbound* was written "upon the mountainous ruins of the Baths of Caracalla, among the flowery glades" and facing "dizzy arches suspended in the air."[5] Any *terrestrial* would see columns here: an *aerial* can only see "arches *suspended in the air.*" More accurately still, it is not the *shape* of the arches that Shelley looks at, but, if we may be allowed to use the word, their *dizziness*. The native land where Shelley dwells, with heart and soul, is an aerial realm, the land of the highest places. The drama of this land of his comes from this very dizziness, from this vertigo which is provoked so that it can be overcome and our victory savored. Thus, man pulls at his chains in order to know the momentum of his eventual liberation. Yet let us make no mistake about it, the positive operation here is liberation. It is this liberation that makes

---

5. *Prometheus Unbound*, Preface.

clear the primacy of the intuition of air, which comes before the solid, terrestrial intuition of the chain. The real meaning of Promethean dynamism is to be found in this vanquished vertigo, in liberty's shaking chains.

In fact, even in the preface to *Prometheus Unbound*, Shelley makes it perfectly clear that we must interpret his Promethean images in a strictly psychological sense:

> The imagery which I have employed will be found, in many instances, to have been drawn from the operations of the human mind, or from those external actions by which they are expressed. This is unusual in modern poetry, although Dante and Shakespeare are full of instances of the same kind: Dante indeed more than any other poet, and with greater success.

Thus, *Prometheus Unbound* is placed under the aegis of Dante, that most verticalizing of all poets, exploring as he does the two verticals of Heaven and Hell. For Shelley, every image is an *operation,* an operation of the human mind; the image has a spiritual, inner principle, even though it is thought to be the mere reflection of the external world. So, when Shelley tells us that "poetry is a mimetic art," we must understand that poetry imitates what it does not see: human life in its innermost depths. It imitates forces rather than movements. The life we see and the movements we make can be described quite adequately in prose. Poetry alone can bring to light the hidden forces of our mental and spiritual life. Poetry is, in Schopenhauer's sense of the word, the phenomenon of these psychic forces. Any truly poetic image will have something about it that makes it resemble a *mental operation*. Understanding a poet in Shelley's sense does not therefore entail a kind of Condillacian analysis of the "operations of the human mind," as a hasty reading of the preface to *Prometheus Unbound* would lead us to think. The poet's task is to set images in motion with his light touch, and so ascertain that in them, the human mind is operating humanly, that these are human images, humanizing cosmic forces. We are led, then, to the cosmology of the human. Instead of living out naïve anthropomorphism, man is restored to profound and fundamental forces.

Now, mental life is characterized by its predominant operation: it desires to grow, to rise up. Its instinct is to seek the *heights*. Thus, for Shelley, poetic images are all operators, *elevation operators,* in fact. To put it another way, poetic images are operations of the human mind in so far as they give us lightness and height, and raise us up. They have only one reference axis: the vertical axis. They are essentially aerial. If a single image in a poem fails to fulfill this function of conferring lightness, then the poem is brought to the ground, and man returns to slavery, bruised by his chains. With all the spontaneity of genius, Shelley's poetics always manages to avoid any such accidental heaviness, composing a sweet, harmonious nosegay of all the flowers of ascension. Shelley can, it seems, with one careful finger, measure the force which lies in every bud and frond, and raise it above the earth. Reading him, we understand Masson-Oursel's perceptive words: "The summits of mental life are very much like tactisms."[6] We *touch* the growing heights. It is in this region of the summits of mental life that Shelley's dynamic images operate.

It is not hard to understand that images which are so clearly polarized in the direction of height can easily acquire social, moral, and Promethean values. Yet these values are not sought after, they are not something that the poet is aiming at. Coming before any kind of social metaphor, the dynamic image reveals that it is a primary psychic value. Love for mankind sets us above our own being and offers no more than a little further assistance to one whose constant desire is always to live above his own being, at the summit of being. Thus, imaginary levitation is very ready to receive all the metaphors of human greatness; however, the psychic realism of levitation has its own driving force, which is, in effect, internal. This is indeed the dynamic realism of an aerial psyche.

L'Air et les songes: essai sur l'imagination du mouvement (Paris: Corti, 1943), 49–53.

Extract III

Virginia Woolf's *Orlando*: The Image of the Tree

*M*en cling to the *certainty* of a hard object with tenacity and passion, and we shall perhaps best understand this if we now look at a dreamer who

6. Bachelard's footnote: Paul Masson-Oursel, *Le Fait métaphysique* (Paris: Presses Universitaires de France, 1941), 49. Note amended. My translation. This book has not been translated.

finds in the company of an immutable tree the strength and firmness of his own being. This, then, is our interpretation of a very fine passage in Virginia Woolf's *Orlando:*

> He sighed profoundly, and flung himself—there was a passion in his movements which deserves the word—on the earth at the foot of the oak tree. He loved . . . to feel the earth's spine beneath him; for such he took the hard root of the oak tree to be; or, for image followed image, it was the back of a great horse that he was riding; or the deck of a tumbling ship—it was anything indeed, so long as it was hard, for he felt the need of something which he could attach his floating heart to . . .[1]

How well the writer suggests this communion of hard objects encircling a core of hardness! The oak tree, the horse, and the ship become one, in spite of their disparity of form, in spite of their lack of any common characteristic or meaning for the eye or the conscious mind. *Hardness* exercises great power over the material imagination, holding sway like some imperialist; it is because of this that images of hardness develop far and wide, going from the solid little knoll where the oak tree grows, to the plain where the horse gallops, then to the sea where, on the ship's deck, we find that all solidity has taken refuge. It is *material comprehension,* the absolute comprehension of images of hardness, that underlies this extravagant *extension* which no logician could ever justify. Indeed, what characterizes primary material images—and hardness is one of these—is their readiness to take on the most varied shapes. Matter is a center of dreams.

Furthermore, a detailed study of this passage from Virginia Woolf's novel will furnish a good example of the two ways in which images can develop, according to whether they go like concepts from one thing to the next or, with a very different movement, live the life of one particular being in all its fullness.

Virginia Woolf follows this second path, and after returning to the image of the hard tree trunk with which she began, she presents us with all that the Tree means for her imagination. Orlando feels his heart grow peaceful as he leans against the oak's hard, firm trunk; he shares in the peace-giving qualities of the quiet tree, the tree that brings quiet to the whole landscape. Does not the oak tree still even the passing cloud?

---

1. *Orlando: A Biography* (London: Hogarth Press, 1928), 20. Bachelard quotes from the French translation.

... the little leaves hung, the deer stopped; the pale summer clouds stayed; his limbs grew heavy on the ground; and he lay so still that by degrees the deer stepped nearer and the rooks wheeled round him and the swallows dipped and circled and the dragon-flies shot past, as if all the fertility and amorous activity of a summer's evening were woven web-like about his body.[2]

Thus, the dreamer has gained from the tree's solidity[3] as it stands in the plain where the ripe corn ripples; the strong trunk and the hard root are the unchanging center, around which a landscape is organized and the canvas woven for a literary picture, for a world in words. Orlando's oak tree is indeed a character *in Virginia Woolf's novel, as the illustrations suggest. If its role is to be properly understood, the reader must have loved, at least once in his life, some truly majestic tree, and felt the influence upon him of this mentor in solidity.*

Last, we should like to propose this passage from the English novelist as a model of material psychoanalysis, of psychoanalysis through images. The tree here is a great oak, a hard oak, great because it is hard. It represents the greatness and the destiny of this courageous hardness. However hard the oak tree's root, the tree will nevertheless carry high into its rough, rustling leaves the being who dreams of its hardness. To this dreamer, who stands quite still on the ground below, the tree will give the movement of birds and sky. In this new example of a reverie that is moored, the dreamer moors his irresolute heart to the heart of the tree, but the tree bears him away on the slow, sure movements of its own life. The dreamer living the tree's deep, inner hardness will suddenly realize that *it is not for nothing that the tree is hard,* as all too often is the human heart. The tree is hard so that it may bear aloft its aerial crown and its winged foliage. It offers men the great image of legitimate pride. Its image psychoanalyzes all pointless and sullen hardness, restoring us to the peacefulness that belongs to strength and solidity.

La Terre et les rêveries de la volonté: essai sur l'imagination des forces (Paris: Corti, 1948), 68–70.

---

2. Ibid., 21.
3. Bachelard's footnote: we find the following line in Laprade: "The oak tree has its rest, and man his freedom." My translation. Victor Richard de Laprade (1812–83) was a minor Romantic poet.

Extract IV
Earth, Fire, and Water: Images of the Smithy

Take Elias Lönnrot's *Kalevala*, for example, which can indeed be described as an epic poem whose subject is work, and it will be perfectly clear that utilitarian words and phrases express only one aspect of the worker's psyche. When iron is tempered in the *Kalevala*, all the powers of the cosmos are bestowed upon it. Let us now look at what characterizes dreams of this kind, which go far beyond the bounds of reality. We read:

> Still there was a trifle wanting
> And the soft Iron still defective,
> For the tongue of Iron had hissed not,
> And its mouth of steel was formed not,
> For the Iron was not yet hardened,
> Nor with water had been tempered.
> Then the smith, e'en Ilmarinen,
> Pondered over what was needed,
> Mixed a small supply of ashes,
> And some lye he added to it,
> To the blue steel's smelting mixture,
> For the tempering of the Iron.
> With his tongue he tried the liquid,
> Tasted it if it would please him,
> And he spoke the words which follow:
> "Even yet it does not please me
> For the blue steel's smelting mixture,
> And perfecting of the Iron."[1]

The smith now needs powers that lie in places far away, powers more sweet than these. So Ilmarinen, the smith of Finnish legend, commands a bumblebee to go in search of hydromel and bring it from "six flowerets... from seven tall grass-stems."[2] *However, a hornet hears his request, and*

---

1. Elias Lönnrot, *Kalevala: The Land of Heroes*, translated from the original Finnish by W. F. Kirby (London: J. M. Dent; New York: E. P. Dutton, 1907), 83–84.
2. Ibid., 84.

instead of this quintessence of the heavens, it brings him black poisons. The hornet

> Brought the hissing of the serpents,
> And of snakes the dusky venom,
> And of ants he brought the acid,
> And of toads the hidden poison,
> That the steel might thus be poisoned,
> In the tempering of the Iron.[3]

In this way the substances of evil are intermingled with the substances of good in order to explain the ambivalence of iron, giving as it does to man both tool and sword. And indeed the poem will end with the exploits of bloodthirsty steel.

Thus, we cannot possibly fail to see that in the beginning the tempering process involves values which are completely foreign to any concern with mere usefulness. Moreover, we should pose the problem incorrectly were we to refer to magic here. Such associations with magical practices do exist, and the relationship between the magical and the technical has quite rightly been studied. There is however a distinction to be made between the oneiric—which is what we are talking about here—and the magical. The oneiric belongs to a rather vague and ill-defined area, to what is in fact the realm of the imagination of matter, the realm too of the oneirism of work.

The imagination of matter worked in the smithy will follow wherever oneiric recurrence leads, and go far beyond the dynamic images of tempering. When the smith sprinkles his fire with water to make it burn more brightly, he is already lost in a deep, material reverie. He well knows that too much water or too heavy a spray would douse the fire. He therefore handles the sprinkler with due care and moderation. What he brings to the smithy fire is, in fact, a kind of *dew*, a beneficent dew possessed of all its secret values. It is no surprise, then, that this practice of sprinkling water on fire should give rise to metaphors which medicine uses in recommending that the fire of life be revived by very careful ablutions.[4]

In certain kinds of reverie, the smithy brings into being a kind of ma-

---

3. Ibid.
4. Bachelard's footnote: cf. Daniel Duncan, *Histoire de l'animal, ou la connaissance du corps animé* (Paris, 1687), preface.

terial equilibrium between fire and water. There is for example a rather odd passage in Goethe where we can see the dialectics of warfare and cooperation in the relationship of fire and water. Let us follow Goethe's material reverie as he contemplates the smithy.[5]

First of all, "the smith softens the iron by stirring the fire which then draws from the metal bar the water it does not need." Thus, the iron bar had imprisoned some kind of residual water, permeated still by the stench and the corruption of the mine. The smithy fire dries this damp iron.

Then comes the second stage in this material reverie: "Once it is purified, the iron is beaten and mastered, and then fed with a strange water, so that its strength is restored." In this way, the tempering process puts water back into the iron, and so it is dreamed as being the *participation* of water in the forged metal.

Goethe's description could well be used as a test which would enable us to distinguish between values belonging on the one hand to oneiric explanation and on the other to rational explanation. This would, of course, be rather disconcerting for a critic whose aim is to reduce images to perceptions. We could make a detailed study of all the colors seen in the smithy or describe the blacksmith's every action, and yet never find anything in all this that explains the material interplay of fire and water.

The critic who, on the other hand, follows the poet to the material center of his images will not be surprised that such deep dreaming should lead to a moral lesson. Goethe's description ends, in fact, with a reference to the familiar image of a man who is "formed" by a master. This moral image will cease to be so trite if we think about the different kinds of participation to which we have been drawing attention here. We could, moreover, read this particular passage in two ways, so that instead of beginning with the image, we could begin with the moral lesson. Goethe's image would then be seen as a *moral* contemplation of work. If no exchange of this kind were made between aesthetic and moral values, Goethe's description would, of course, remain inert.

Since reveries born of tempering are so numerous and so very free, perhaps we might be allowed to take up in our own way those dreams of the avarice of substance which, in our view, accompany tempering and its reveries. The hot iron is *enriched* by all the powers of fire: are we just to

---

5. Goethe, *Maximen und Reflexionen*. My translation here. Bachelard quotes from the French translation.

let it *lose* its fire and heat? No indeed, for by plunging it suddenly into icy water, we are dreaming of a process that *blocks* all the virtues of fire within its substance. We have only to dream substantially and give ourselves with all our material imagination to dreams of the riches of substance, to strong dreams, and we shall then have the *idea* of shutting fire into iron by means of cold water, of shutting the wild beast, fire, into its steel prison. When Wagner describes the tempering of Siegfried's legendary sword, *Notung*, he writes:

> In the water flowed
> a fiery flood:
> fury and hate
> hissed from the blade!
> That fire was soon quenched
> by the fiery flood;
> no more it stirs.[6]

Mastery of something does not mean destroying it, but rather caging it. In the *Kalevala*, too, we read that in the tempering trough,

> Then indeed the steel was angry,
> And the Iron was seized with fury.[7]

And in alchemy we find many similar images of *trapped fire*, of enclosed fire. Tempering by the blocking of fire is a normal dream. It is one of the smithy's dreams.

All the lyric qualities of the *shining sword* now come together. The sword does not simply reflect the rays of the sun. On the battlefield, whenever the sword is struck, the fire imprisoned within it will leap forth. The sword will shine not because it reflects light, but because of its own inner virtue, not because of some violent blow, but because of its own valiant metal. The "heat" of battle is already latent in the "fire" of the sword which has been so heroically tempered. The sword that is forged and tempered with all the smithy's dreams is heroism made material. It is legendary in its very substance, before it ever belongs to the hero.

6. Richard Wagner, *Siegfried*, Act 1, Scene 3, in *The Ring of the Nibelung*, trans. Andrew Porter (London: Faber Music, in association with Faber and Faber, 1977), 186. Bachelard quotes in French.
7. *Kalevala* 85.

Reverie will of course cease when work is done, and we shall see that the tempered steel has become harder, that it can cut into the iron which has now indolently grown cold. Yet these objective experiences simply confirm our first reveries. Thus, reveries are in fact oneiric hypotheses which, were we to look for them, we should discover at the root of even the clearest of technical processes.

These, then, are the profound, deep-rooted reveries which must be awakened if we are to give moral images their full force, or more accurately, if we are to give morality the full force of images. A well-tempered character can be so only in adverse circumstances, these being both explicit and abundant, for we must understand that tempering is a struggle, that it triumphs in the battle of the elements, in the depths of substance itself. Etymology can give us only meanings which have no virtue, meanings which are quite simply nominalist. The realist value of words is to be found in our first reveries, and nowhere else.

La Terre et les rêveries de la volonté: essai sur l'imagination des forces (Paris: Corti, 1948), 149–54.

"The *realist value* of words," to quote the last phrase in the last of these four extracts, "is to be found in our first reveries, and nowhere else." This seems an odd and contradictory way to describe language. However, the context of this statement clarifies Bachelard's intention, which is to shock us out of our preconception that words refer to things, that their meaning is fixed by history and social usage. Bachelard provides abundant evidence that words are not just labels, as Bergson thought, indicating "kinds of things" and the ways they are used[1] (Bergson 1964:117 [1913:153]). Even where the descriptions of the practical, technical process of tempering are concerned, words like *purified* and *beaten* are ambivalent, describing the facts of steel making, while at the same time expressing values that are "moral" and "human." He does not, though, argue that the meaning of words is purely subjective: "the *realist value* of words is to be found in our first reveries," and these are material reveries, in Bachelard's view, reveries which in writing about Virginia Woolf he describes as "moored." Material reverie explores the possibilities of matter and is itself made possible by those possibilities—it is the oak tree's hardness that is the starting point and mooring cable of Virginia Woolf's reverie—so that these "written reveries" maintain the tension of man and matter, subject and object, at the same time resolving it.

Bachelard's main concern is the practice rather than the theory of language: he is a reader who writes, "continuing" what he reads through his own use of language, and because of this, experiencing "openness." This may appear a wayward, even simple view of reading, yet as so often in Bachelard, it conceals something more complex. When he distinguishes between the activity of literary criticism and what he himself does, when he censures literary critics for their reductive reading which seeks to control and limit language, centering and fixing it, what is at stake is less a conception of language than of the human being. "We have only to read a true poet," he declares, "to understand that language reveals us to ourselves" (DR 184). How though can poetic language, with its manifest oddity, even opacity, "reveal us to ourselves"? If we go back to Bachelard's phrase, we see that this revelation is the work not of poetic language in itself but of reading. When we read, we discover the possibility of a different language from the one we use in our everyday lives, and this consciousness of new language is not passive, Bachelard believes; it changes our language, it changes us. It is in this way that the reader is made different by difference, that he performs a linguistic break, and is unfixed by language.

This consciousness of written language involves more than consciousness of the intrinsic difference of language: it is consciousness of difference that someone else has made. The reader must beware of attending passively to this newfound other language, of letting "consciousness of language" become fixity, because this would deny and destroy the dynamic structure of consciousness, the interdependence of subject and object that the simple, accessible experience of reading a poem shows to be a fact, not just a theory. The best way to ensure that we continue to be transformed by language is to write what we read, to unfix our own language actively in response to the poet's language. To read a poem "pen in hand" is to discover ourselves as subjects who are conscious of being transformed by an object, by difference, who are in turn conscious of our own possibilities, of difference in us, of openness. Reading not only reveals what we are, it restores us to ourselves. It has, for Bachelard, an ontological dimension; it is far more than just a pleasant pastime or a way of acquiring new ideas. Without the desire to read, we cease to be human, for we lack that essential, sustaining relationship with an unfixing object.

Chapter 8

Applied Rationalism: 1949–1953

> All thought reveals the thinking being, and a chemical analysis is *also* an analysis of thought ... A complex psychology necessarily accompanies a complex science.
> —*Le Matérialisme rationnel*

Immediately after these books on poetry, Bachelard published three further works on the philosophy of modern science, *Le Rationalisme appliqué* (1949), *L'Activité rationaliste de la physique contemporaine* (1951), and *Le Matérialisme rationnel* (1953). His return to epistemology is almost as unexpected as was his defection some ten years earlier. Since his appointment to the chair in the history and philosophy of science at the Sorbonne in 1940, he had published only one book on science, *La Philosophie du non* (1940), together with eight book reviews or short articles and papers on epistemological matters. This limited scientific activity can obviously be explained by the war. The war, however, was not the reason for his work on poetry, which was plainly not an opiate, and given the value of poetry to him, it would also seem oversimple to account for his return to epistemology by the ending of the war alone. Bachelard says very little about this period—he was not in any case given to revelations about his personal life—but there are two instances when he does refer to it, in one directly, in the other indirectly, which suggest why at the age of sixty-five and on the verge of retirement he should begin to produce a new series of books on science. Both references are made a few years after the war, but both ring very true.

Bachelard writes of the "suffering" and "trials" of this time when recalling his friendship with Jean Cavaillès in his 1950 postface to Gabrielle Ferrière's book *Jean Cavaillès, philosophe et combattant (1903–1944)* (*Eng.*

rat., 178–90). Cavaillès had given him hope, by his visits and, much more important, by working and continuing to work when a prisoner of war—he died in Nazi hands in 1944—writing a major book on the philosophy of modern mathematics, *Sur la logique et la théorie de la science*, published posthumously in 1947. Cavaillès, Bachelard says, was not just a war hero, he was a "great mind" (179), and this great-mindedness, founded as it was on his sense of the importance of science not as something useful but as a "human creation" (189), seems to restore Bachelard's faith in his own epistemological work. Yet it could also be argued that this war had shown the destructiveness of science. Bachelard does face up to this problem in a conference paper given at Geneva in 1952, "La Vocation scientifique et l'âme humaine," and this is the second, though indirect, reference to the war years. It has become common to blame science, he says, for unleashing great forces upon us, but it is human beings who have done this, and who now confront a "moral problem" (11, 28). This moral problem is not whether we should or should not go on with science: modern man cannot opt out, he is "situated in science" (11, 17). The moral problem concerns the transmission of knowledge, which should be seen not as "power" but simply and solely as knowledge (12). The problem is not how to choose between the good or evil uses of science, but how to teach the "human value" of science (29), how to make people understand that human nature and indeed the very structure of our consciousness make scientific progress necessary and inevitable, that science is our vocation because of our human being. It is surely significant that both these references to wartime suffering and destruction lead Bachelard to the same conclusion, and that this certainty of the importance of "the human factor in human science" (*ARPC* 301–2) runs through these three books. It would seem to be the experience of the inhumanity of war that, paradoxically, obliges Bachelard to reconsider and reassert the humanity of modern science.

Has Bachelard anything new to say in 1949? The phrase "applied rationalism" might suggest a restatement of old themes, clearly expounded in *Le Nouvel Esprit scientifique* some fifteen years before, and already implicit in his *Essai sur la connaissance approchée* in 1928. He used this phrase for the first time in *La Philosophie du non* (1940) in order to underline the difference between the closed, a priori reason of traditional philosophy and reason in modern science, which, because it is always applied, not only reaches beyond itself but in doing so modifies itself, which is therefore polemical and open. However, he did not develop the idea in 1940—the

Applied Rationalism

phrase is used only once (6)—since what interested Bachelard was less the relationship between reason and reality than what he called "the reaction of scientific knowledge on the structure of the mind" (7). He wished to shake the philosopher's acceptance of the permanence and identity of the *cogito*. Because of this, he took as his theme the phrase the "philosophy of no," suggesting not the polemics of reason and reality but rather that of the mind with its own past, an argumentative state of mind which produces "a profound human mutation" (144). We know how important this idea is to Bachelard, and it might seem that "applied rationalism" signals a regression, a turning away from this concern with the psychology of the scientific mind, the reintroduction of an idea which earlier proved not fruitful.

First impressions inevitably mislead, and these two extracts from *Le Rationalisme appliqué* have been chosen to show that far from merely refurbishing old ideas, Bachelard now understands and uses the phrase quite differently. This new conception of "applied rationalism" marks in fact a new conception of the rational subject. These three books are an enquiry into modern science as a culture and a human activity, conducted for the first time against Husserlian phenomenology in particular and the philosophies of existence in general. They present a sustained attack on idealism in its most recent and beguiling form. Time and again they refute the notions of "the imperialism of the subject," of "the solitary consciousness" (*RA* 8), and with them, as the last sentence in Bachelard's last book on science firmly declares "the rationalism of identity" (*MR* 224). The nub of Bachelard's argument is not simply the dialectical relationship between reason and reality, "applied rationalism" in the sense of reason being polarized by reality, but the dialogue of mathematician and experimental scientist, "applied rationalism" in the sense of "the *application* of one mind on another" (*RA* 12). The first passage, "Cogitamus: On the Psychology of Coexistence," from chapter 3, "Rationalism and Co-rationalism," examines the relationship between "a rationalist *I* and *Thou*." The second, "The Divided Subject," from chapter 4, "Intellectual Self-Surveillance," suggests how this relationship is internalized by an individual subject.

The idea of the *cogitamus* is new in Bachelard. Until now, he has always considered the scientist as a *cogito,* a subject whose formative relationship is with the object of his thought, as a lone figure, working in his laboratory without companions (see Chap. 3, Extract II).[1] Indeed, companions, relationships with other people, were seen in *La Formation de l'esprit scien-*

tifique as obstacles to scientific knowledge that must be "psychoanalyzed."[2] In his books on poetry too, what counted was the relationship between man and matter, and he was always critical of Freudian psychoanalysis for the importance it gave to relationships between people. Bachelard will continue to discuss modern science in terms of a dialectical relationship between reason and reality, which in fact underlies his new idea of "regional rationalism," put forward in *Le Rationalisme appliqué* and illustrated in *L'Activité rationaliste de la physique contemporaine* and *Le Matérialisme rationnel*. Rationalism in modern science is "fragmented" (*RA* 131), its different regions determined not by immediate experience but by *"reflection"* (122), by "noumenal experimentation of phenomena" (124), so that, for example, the phenomena of both osmotic and gas pressure belong, despite their difference in immediate experience, to one "rational region," that of pressure (125–30). He goes on to examine the different regions of reason in modern science: "electrical rationalism" (*RA*, chap. 8), "mechanical rationalism" (chap. 9), the "rationalism of energy" (*ARPC*, chap. 5) and of quantum mechanics in general (chaps. 3–4, 6–10), "material rationalism" in modern chemistry (*MR*, chaps. 3–5), the "rationalism of energy" in chemistry (chap. 6), and the "rationalism of color" (chap. 7). Reason in twentieth-century science is shown once again to be a differentiating activity. Bachelard also considers, as he discusses regional rationalism, the effect of knowledge on the knower. His idea of a non-Cartesian *cogito* now seems too simple. If, as he argues, modern science requires *"bi-certainty"* (*RA* 3), that is to say the agreement of mathematician and experimental scientist, then an individual scientist must, when trying to solve a problem, envisage the solutions that other scientists would offer. A mathematician, for example, must not only envisage applications by experimental scientists after solving a problem, but also be certain before he solves it in terms of his own frame of reference that it can also be solved by others who tackle it from both the same and very different points of view. As a result, his consciousness of a problem is mediated by his consciousness of the problem-for-others; the *cogito* is founded on the *cogitamus,* on the dialogue of the mathematician and the experimental scientist, "applied rationalism" in a new sense. It is a complex idea whose implications will best be unraveled by following the detail of Bachelard's argument in these two extracts.

The first extract begins by looking at science as a process of posing and solving problems. These problems are neither disparate nor posed at

Applied Rationalism

random; they are part of a structure, questions asked because of what is already known and which change what is known. An individual poses a problem against the ground of other people's thought, which precedes his own and which he also shares with others, his colleagues and contemporaries; hence the notion of "the union of intellects." Problems in modern science are not therefore questions that simply occur to individuals, and in the same way, solutions to those problems require "rational communion," the cooperation of a "rationalist *I* and *Thou*." It is not just the difficulty of modern science that imposes the need for teamwork, though this is a factor Bachelard takes into account, but much more the structure of consciousness. Implicit here is Bachelard's long-established criticism of the Cartesian *cogito:* I am not simply a being who thinks, but one who thinks about a problem, who is therefore consciousness of a problem. This would seem to be very close to Husserl's conception of the intentional structure of consciousness—consciousness of something other than itself—but it is important to note that Bachelard adapts this, referring early in *Le Rationalisme appliqué* to "the intentionality of applied rationalism" (10). The problem to be solved "polarizes" consciousness; it determines the way we think, it restructures past knowledge, it redirects our thinking, and in a word, it changes us. Consciousness is more than "consciousness of a problem"; it is "consciousness of being changed by a problem," of the "mobilization" of our intellect. Moreover, because science now requires "bi-certainty"—dialectics, "applied rationalism"—this consciousness of a problem is also consciousness of other people, of a problem-for-others: my solution to a problem must also be the solution others find. The *cogitamus* therefore in fact precedes the *cogito;* it is "the fundamental *cogito* of the rationalist subject." Without this "*cogito* of mutual obligation," this "*cogito* of obligatory mutual induction," my own thought is in doubt: *cogitamus ergo sum*. Echoing an equally famous phrase, Bachelard points to the consequence of this, namely that "coexistence precedes existence."

This is far more than just a humorous remark, a playful cuff aimed at an influential and fashionable philosopher. If we recall that earlier in this passage Bachelard referred to a "thinking coexistence," we realize that he is taking up position not only against Sartre but against all the philosophers of existence. His objections center on the failure of these philosophies to perform the epistemological break, to attend to the mode of existence of man and things in twentieth-century science. Instead, they remain philosophies of lived experience, condemned therefore to half truths. For

Bachelard, modern man is *situated* in science, and this situation is not, as it was for Sartre, a datum; it is "rectified reality" (*MR* 198), in which objects are not "given," not "found," not "natural," but artificial, constructed by man (*ARPC* 9). Sartre's "human reality" is in a world of tables, chairs, cups, and bicycles that get punctures; Bachelard's is in one of electrons and neutrons, of telephones and electric lights. What is the distinction? All are human artifacts, but Sartre's are the work of *homo faber,* Bachelard's of *homo mathematicus,* of *homo aleator;* they are first and foremost "objects of thought" (*RA* 63). From a scientific point of view, Sartre's analysis of things is simplistic: this cup is real, he declares, because it *is* there and it *is not* me (1943:13 [1957:xlvii]); it is as it appears to me. Bachelard is scathing about this reduction of reality to appearance, he considers Sartre's insistence that appearance is reality as placing him fairly and squarely in the nineteenth century. "Like it or not," he says, "the flux of electrons, presented as the deep reality of the electric current, is an example of being that is beneath appearance" (*ARPC* 129–30). Science now makes it possible for us to speak of *being:* "energy *is*. It *is* absolutely . . . being is energy— and energy is being. Matter is energy . . . There is nothing *behind* energy" (*MR* 177). Bachelard believes that we are all of us in science. Our situation is therefore not contingent, nor is it absurd or gratuitous, but it is a situation we choose, that we produce, that we order and constantly reorder. Consequently, our situation is more than a "human reality"; it is a "social reality" (*RA* 6), it is for-us (*MR* 198). A philosophy that seeks to pay attention to the facts must, in Bachelard's view, attend to the facts of modern science, and he is critical of Husserlian phenomenology in particular for failing to do so (*ARPC* 7).

Coexistence is for Bachelard a fact, and from it he develops the idea of what he terms "the divided subject," or more accurately, "the *divided thinking subject*" (*RA* 64), which this second extract explores. He begins by considering "reflection," thinking in general, and argues that it is an activity controlled by our awareness of other people. We "internalize" others, as it were, and our intellectual progress depends on this. Thinking involves more than a dialogue between subject and object; it requires a dialogue, an argument with someone else. A hypothesis, or as Bachelard calls it here a "fiction," "fictitious thoughts," is in effect another point of view, a different way of thinking about a problem. When I think, I consciously wear a mask, I conceal my identity and take on a different one: the mathematician must think as if he were an experimental scientist, and vice versa;

he must hide from himself his own rational past if he is to become more rational. Bachelard insists that this process of self-division must be conscious, controlled, deliberate. What he goes on to call "self-surveillance" is not to be confused with the Freudian superego, and he distinguishes carefully between the two ideas (*RA* 69–77). The superego internalizes parents or some other authoritarian figure from our childhood; it is therefore dogmatic, historical, and entirely *closed*. There is no dialogue here; the internalized authority figure dominates, the judge is never judged, and the two poles of the unconscious personality remain far apart. Intellectual self-surveillance, on the other hand, judges the past; it is a process of rectification in which the roles are constantly reversed, the judge is swiftly judged, and the poles of the divided self are held close together so that their relationship is dialectical and open.

The divided, thinking subject must maintain and sustain its divisions through a rigorous process of self-surveillance. "Simple surveillance" is essentially an attitude of empiricist thought; it is consciousness of the "contingency of facts," and at the same time, of our freedom to think about them. It marks the first break with ordinary experience, with obedience to facts, and it is self-surveillance precisely because this break is made. (Surveillance)^2 is in fact applied rationalism, "consciousness of the rigorous application of a method," the dialectic of reason and experiment, demanding therefore constant breaks with both reason and experiment. More rigorous and fragmented is the divided consciousness of (surveillance)^3, for it breaks with the rational past, with method itself, and with the rules of reason. Last, Bachelard considers the possibility of (surveillance)^4, suggesting that it lies beyond science in poetry. Reading poetry, we master thought itself; we are conscious of ourselves breaking with thought and with life, conscious too of the need to maintain this break. (Surveillance)^4, though difficult to grasp, is a state of extreme self-consciousness, consciousness of a divided, fragmentary, momentary self. It is seldom attained, and most of us must be content with a lesser degree of self-surveillance. What we learn about ourselves is the same, namely that we must respect and also exploit "the mind's dialectical powers, and make the *divided subject* conscious of its divisions, desirous of dividing itself as it divides."

Bachelard's reference in this second extract to *La Dialectique de la durée*, and specifically to his chapter on temporal superimposition, with its description of "separate, consecutive existences," is important because it points up the change in his thinking. There is a difference between, for example, (cogito)³ and (surveillance)³ because self-surveillance implies the *cogitamus*, because coexistence both precedes existence and is the condition of "superexistence," of a "hierarchical existence" (*RA* 65). The scientist is no longer regarded as a solitary consciousness; indeed there is no evidence for Jacques Gagey's assertion that in *Le Rationalisme appliqué*, "the will to rationality isolates" (1969: 190). Applied rationalism here means thinking with others, consciousness of others, any desire to be "original," "unique," and "absolute" being, as this first extract shows, in Bachelard's eyes foolhardy. He is now very much aware of what he calls the "socialization of truth," "the union of minds in the truth," stressing in all three books that modern science means a community, a culture in which we are all of us situated, and from which we must all learn to benefit: we discover our psychological complexity, divided selves bound to others, and with them to that rectified reality inseparable from scientific reason.

Extract I
Cogitamus: On the Psychology of Coexistence

In order to understand the terms of a problem, we have to normalize all the questions related to it; in other words, we have to develop a kind of topology of its problematics. Any aberrant question ought to be eliminated, of course, and we must establish a clearly defined set of problems. Time and time again, textbooks will tell us that a well-posed problem is a half-solved problem. Karl Marx has said even more succinctly that in formulating a question, we are in fact answering it.¹ Let us take this to mean that if we ask intelligent beings an intelligent question, we bring about the union of intellects.

However, rather more is needed here than the union resulting from the establishment of a well-defined problematics. We need to see that as we pass from the problem to its solution, there comes into being what

---

1. Bachelard's footnote refers to the French translation of Marx. The English translation of the work referred to is as follows: Karl Marx, *A World without Jews*, trans. Dagobert D. Runes (New York: Wisdom Library, 1959). Bachelard has in mind this phrase from chapter 1, "The Jewish Question": "How does Bauer solve the Jewish question? His formulation of the question itself contains his solution" (3).

philosophers of micro-epistemology might well call an atom of rational communion.

Let us try to determine the complex structure of this atom of rationality by examining the way relations are established between a rationalist *I* and *Thou*, when each strives to help the other toward the rational solution of a problem.

First, the object must be made the subject of a problem, the subject of the *cogito* being consciousness of a problem. Thus, the thinking being thinks at the very limits of his knowledge, having first enumerated all that he knows and that might enable him to solve the problem before him. Consequently, this enumeration, this consciousness of a dynamic order in ideas, is polarized by the problem to be solved. In the rationalism we are taught, enumeration is codified; it is restricted to operating in one clearly defined direction only, it is fixed and limited by its bases. However, in the rationalism that asks questions, it is these very bases that are put to the test, that are called into question by the question we have asked. The Problem is the active summit of research. And so in foundation, coherence, dialectics, and problem we have all the elements of rational enumeration, all the moments of this mobilization of the intellect.

It is here, in the explicit development of these four moments of applied rationalism, that the *cogitamus* comes into being, uniting the rationalist *I* and *Thou* in one single thought and consequently in one thinking coexistence. By means of this *cogitamus*, the *I* and the *Thou* can be culturally superimposed on one another in exactly the same way that two figures are said by mathematicians to be *superimposed and congruent*. Two rationalist minds do not need to be completely identical in order to become conscious that they are concordant; they have only each to assume the role of objectively controlled thought. It is these controlled roles, these functions functioning with respect to a normalized object that offer the best ways of developing *discursive agreement*. In other words, the rational *cogitamus* is consciousness not so much of a common *property* as of a common *income*. It tells us our thinking will be fruitful. It creates an obligation to think in agreement with one another; it is, in short, our common consciousness of apodictic knowledge.

If we wish to formulate the fundamental *cogito* of the rationalist subject, we must select from among the formulae of interpsychology those that correspond to an induction which is absolutely and entirely *certain*. The rationalist subject develops in the *certainty* that it will be possible for him to teach, and this must perforce involve a rationalist other. Once this certainty has been achieved and a degree of psychological insight gained in the process through preliminary psychoanalysis, the rationalist subject can anticipate the different ways irrationalism will resist. He may even employ a slightly demoniac psychoanalysis and enjoy watching his adversary think, fated to remain in error because of his attachment to irrational values. The behavior patterns characteristic of irrational singularity are psychoanalytically fairly clear. The different ways of seeking originality can be easily classified. Faced with a thinker who claims he is an absolute being, rationalist psychoanalysts can say to themselves: we, the several, watch him feigning uniqueness.

In these circumstances, it seems to us that the *cogito* of mutual obligation should be expressed in its simplest form as follows: I think that you are going to think what I have just thought, if I inform you of that rational event which has just obliged me to take a step further and think beyond what I was thinking. This, then, is the *cogito* of obligatory mutual induction. Moreover, this rationalist *cogito* is not strictly speaking of the order of interconclusion.[2] It is formed before the *I* and the *Thou* agree, since in its first form it appears in the solitary subject, as a certainty of agreement with a rationalist other, once the pedagogical preliminaries have been established. We can *oblige* others to draw conclusions: since I recognize that what I have just thought is something that is entirely normal for normal thought, I have the means by which I can force you to think what I think. Indeed, you will think what I have thought in so far as I establish you as being conscious of the problem whose solution I have just found. We shall be united in the proof as soon as we are absolutely sure that we have both quite clearly posed the same problem. Moreover, the solution of a problem brings a new clarity to its terms, and it does this by

---

2. A note on my translation may be helpful here. Bachelard's phrase is "de l'ordre de l'interconstatation," the last word being one he coins, based on the noun "la constatation." The verb "constater" to which this is related is defined in French as "to establish the truth or reality of something by direct experience" (*Petit Robert* dictionary) and is variously translated in English as "to note," "to notice," "to take note of," "to make an observation," "to state a fact," "to make a statement of fact," all of which lose something of the French meaning. In order to convey the full sense of "constatation" in my translation I have used two words in this extract, "observation and conclusion." "Observation" is given as a synonym of "constatation" in French. Used by itself, "observation" could however suggest a purely perceptual process, inappropriate to Bachelard's use of the word here. I therefore attach "conclusion" to it in order to try to convey the French sense of establishing that something is true or real.

Applied Rationalism

recurrence. The relationship between problems and their solutions is of considerable importance epistemologically, and it dominates the empiricism of observation and conclusion. Whether observation and conclusion are psychological or sensory, the moment they take note of the fact that a problem has been solved, they benefit from the *values* of a well-ordered discovery. Here, then, we have the validation of a method, the proof of the efficacy of thought, and the socialization of truth.

Of course, two minds can find themselves united in the selfsame error. Yet the deepening shadow is not simply and solely the inverse dynamics of nascent lucidity. Error *descends* toward convictions, whilst truth *rises* toward *proof*.[3] The discussion that we ought to open at this point would take us back to descending psychology, the study of which can have no place in the psychoanalysis of knowledge until the time comes for us to examine irrationalism and its arguments. However, if we now pose the problem of error from the standpoint of *scientific* error, we shall see very clearly, or rather concretely, that *error* and *truth* are not symmetrical, as a purely logical, formal philosophy would have us think. In science, truths are grouped together in a system, while errors vanish and are lost in an amorphous magma. In other words, truths are linked together apodictically, while errors accumulate assertorically. There is in present-day scientific thinking an obvious disproportion between, on the one hand, truths which are rationally coordinated and codified in books guaranteed by the scientific community, and on the other, a few errors which persist in a few bad books, books that as a rule bear the stamp of loathsome originality.

Consequently, if we base our argument on the pedagogics of scientific thinking and examine scientific culture as it is today, then the concept of *epistemological value* is clear and the nature of the union of minds in the truth is quite unmistakable. It is here in these distinctions, which may seem nice but which are very real, that we shall see differences emerge between the psychologism of observation and conclusion and the psychologism of normalization. The total condemnation of psychologism, which people are so often quick to reject, misunderstands these fine distinctions, small but nevertheless essential.[4]

---

3. Bachelard's footnote: cf. Nietzsche, "that which convinces is not necessarily true on that account: it is *nothing more or less than convincing*. An observation for donkeys." *The Will to Power*, Vol. 1, trans. Anthony M. Ludovici (Edinburgh and London: T. N. Foulis, 1909), 18. Bachelard quotes from the French translation.

4. Bachelard's footnote: Movements of proof which are less determinant than movements of apodictic proof can also be analyzed in terms of dual psychology. Where problems of

How then can I fail to posit the *coexistence* of a common thought when it is from the *Thou* that I have proof of the fruitfulness of my own thought? Along with the solution of *my* problem, the *Thou* brings me the decisive element of my own coherence. He it is who sets in place the keystone of a system of thought I could not complete. Where he and I are concerned, coexistence precedes existence. Coexistence does not just reinforce existence. At least, the reinforcement of existence which a particular subject may receive from another rationalist subject is but one aspect of more clearcut metaphysical distinctions. In the *I-Thou* of rationalist thought, we see control, verification, confirmation, psychoanalysis, teaching, and normativism, all of which are forms of coexistence of varying rigidity. And in our finest hours come the enhancements of apodictic existence, of coexistence through apodicticity.

When we know the support the apodicticity pervading knowledge brings us, then we live the division of our own self, a division that can be described by the two words existence and superexistence. The subject raised to superexistence through the coexistence of two subjects will see setting up within him the dialectics of the controlling and controlled subject. Within his own mind, opposite his *I*, he will set up a kind of vigilant *Thou*. The word "dialectics" is no longer strictly accurate here, because the pole of the assertoric subject and the pole of the apodictic subject obey an obvious hierarchy. The *cogito* that has left the first pole and set itself up as the highly valued subject of a rationalist *cogito* cannot go back to being a *cogito* of observation and conclusion, to an intuitive *cogito*. The *cogitamus* is resolutely discursive. The coexistence of rationalist subjects casts its own net of logical time upon empirical time. It brings order to experience, in-

---

knowledge are concerned, all the help that anyone can give us is welcome, however limited it may be. Edgar Quinet in his book *La Création* writes about the moment in the development of science when the geology of the Alpes de Maurienne threw paleontology into disarray. On this same subject, Lyell said to a colleague: "I believe it because you have seen it; but if I had seen it myself, I should not believe it." This anecdote—so revealing from a psychological point of view, with its unusual nuance of a kind of *humor of courtesy*— also has a certain epistemological significance. It shows that surprise, so useful in scientific culture, cannot remain *individual*. No sooner are we ourselves surprised than we want to surprise someone else. We learn in order to surprise others. If we teach one another, then we surprise one another. This is surely ample proof of the need for renewal that is the driving force in all intellectual development! Even in the smaller cultural areas like geology, with its many subdivisions, some new event will always rouse the scientist from his dogmatic slumbers.

deed it repeats each and every experience, each and every experiment, so as to be sure of its triumph over all contingency.

The *cogitamus* brings coexistence to us in all its intricate complexity.

Le Rationalisme appliqué (1949; Paris: Presses Universitaires de France, 1970), 56–60.

Extract II
The Divided Subject

We have only to consider the normal adolescent, the normal adult in the civilization in which we are now living, for it to become perfectly clear that thinking, as it is usually practiced, is an essentially *secret* activity. Of course, thinking will tend to *reveal* itself, it likes to reveal itself and find expression in many different shapes and forms. Yet when thinking is at its most elaborate, it is most often secret, it is indeed *first and foremost* a secret. Emotion and desire, pain and pleasure, are all directly revealed. They can be read from our facial expressions, and in their elementary forms they escape our control. The very opposite is true of reflection which is by definition in two phases, with its second phase controlling adventitious thought. It is very rare, indeed it is not quite normal, for us to let slip our thinking, to allow it to show, or to put all of it into words.

The dualism of the secret and the revealed, which is in effect an essential dualism, is therefore particularly clear where reflective thought is concerned. It can indeed serve as a sign indicating not just that thinking is well made, but that the thinker has taken full responsibility for it. It is only when this dualism has established its ascendancy that the mind will possess the *freedom to think*. We can think freely only if we have the power to conceal our thinking wholly and completely. The time will soon come when, in order to defend itself against the use of inquisitorial tests, free thought will have to rediscover the hypocrite's talents. We must show that, where intellectuality is concerned, this self-mastery can be attained only by means of a nonpsychologism that goes beyond psychologism, in a kind of freedom to think about thought itself. Yet without some kind of mask, we cannot attain this freedom, and the mere mask of negativism will not do. Here we must stress the importance of fictitious thoughts. If we consider this idea of *fiction* from the functional point of view, we shall find that it is a divisive factor for the subject. This is, of course, because we are talking about a fiction which the subject brings against himself as he pursues his quest for knowledge, experiencing deep within himself the dialectics of objection and answer, of supposition and control. From many points of view, a *larvatus prodeo* plays a kind of inward hide-and-seek with the *cogito*.[1] An extraverted *larvatus prodeo* would lead to formulae such as these: I say that I think, therefore I do not think what I say; I am not what I say I am; I am not entire either in the act of my thinking or in the act of my speaking. The subject who expresses himself is a process of self-division.

Yet the *larvatus prodeo* is so human an activity that it comes to be a determining factor for the thinking being. I am, with regard to my own self, a pretence, a sham. I am a *hypothesis of being*. My ever advancing thought is the advancing of a hypothesis. If this hypothesis is successful, I become intellectually that which I was not. Yet where am I the self that is becoming? Am I recalcitrant thought or recurrent thought? Does not every new thought recreate some kind of past in me, since a new thought is automatically a judgment of a thinking past?

Hence, if we wish to see real thought in action, we must come in the end to an ontology distributed over two or more levels of being.

These divisions will be especially clear when control functions come into play. The more sensitive these are, the more precise the different levels of being established by the division of the subject. Indeed, we cannot appreciate the full importance of these control functions if we restrict ourselves to the differences between what lies hidden and what is expressed, and we shall see that the *doublet* controlling and controlled is active at

---

1. Bachelard's footnote: Descartes, *Complete Works*, Vol. 10, edited by Adam and Tannery (Paris: Léopold Cerf, 1908), 213. Note amended. His reference is to the first of Descartes's *Cogitationes Privatae*, the sentence from which he quotes being as follows: 'ut comoedi, moniti ne in fronte appareat pudor, personam induunt: sic ego, hoc mundi theatrum conscensurus, in quo hactenus spectator existi, larvatus prodeo.' No English translation having been published, my version is the following: "just as actors put on a mask lest shame be seen on their faces, so I, as I mount the stage of this world where till now I have been a spectator, put on my mask, and so walk on to play my part." Bachelard's footnote adds (quoting from the French translation of Nietzsche): Nietzsche distinguishes between men and beasts in these terms: "the beast cannot dissimulate, it conceals nothing." See "The Uses and Abuses of History," *Thoughts Out of Season*, Vol. 2, trans. Adrian Collins (Edinburgh and London: T. N. Foulis, 1909), 7.

Applied Rationalism

all levels of our intellectual and moral culture. We have already seen that rationality is constituted through the dialogue of master and disciple. Now, speaking more generally, we can say that the mind is a school, the soul a confessional. All that is deep and inward is dualized.

Yet here again, we cannot really situate the precise centers of division if we do not first look at the problem where it is at its most confused, at its most indistinct, and its most disguised. Scientific culture can alone establish the mind's dialectical powers, and make the *divided subject* conscious of its division and desirous of dividing itself as it divides. Thus we see that this double consciousness enhances us. Even error has a useful part to play in the progress of knowledge, thanks to rectification . . .

In our efforts to acquire a scientific culture, the function of self-surveillance takes on certain compound forms which show us very clearly the psychic action of rationality. If we study it at all closely, we shall find yet further proof that rationality is specifically secondary in character. Only when we understand that we understand can we be really established in the philosophy of the rational, only when we can point out with unfailing accuracy errors and mere semblances of understanding. In order that self-surveillance be fully assured, it must somehow be itself held under surveillance. Thus, there come into existence forms of *surveillance of surveillance* to which, for the sake of brevity, we shall give the exponential notation $(surveillance)^2$. We shall, moreover, set out the elements of a surveillance of surveillance of surveillance, in other words, of $(surveillance)^3$. It is not difficult to grasp what is meant by an exponential psychology, with respect to this problem of disciplining the mind, and we can readily appreciate the contribution made by this exponential psychology to the ordering of dynamic elements in both experimental and theoretical conviction. The sequence of psychological events is determined by causalities of very different kinds, according to the way they are organized. This sequence cannot be revealed in the continuous time that belongs to life. To explain such varied sequences, we need a hierarchy. This hierarchy cannot but involve the psychoanalysis of whatever is useless and inert, superfluous and ineffectual. In an earlier chapter we emphasized the fact that any apprehension of an object must first eliminate the features of that object which are regarded as having no importance. And this observation is just as valid for the dynamic features of phenomena as it is for the static features of objects. Thus, the phenomenon is now apprehended in hierarchical time, it is understood in a time which, by eliminating conditions that are aberrant, contingent, or accidental, attributes coefficients to a logical order, a rational order. As we examine this mastery of the evolution of phenomena, we shall find here once again the temporal themes already presented in *La Dialectique de la durée*, in particular in the chapter on superimposed time. Once we possess a phenomeno-technique, we see that the temporality of phenomena often develops in accordance with the causality of thought. A physicist's surveillance of his technique is now conducted from the standpoint of the surveillance of his thoughts. His constant need is for *confidence* that his apparatus is working *normally*. He makes sure that its perfect working order is maintained and guaranteed. The same is true with regard to the purely psychic apparatus of correct thought.

Having suggested how complicated the problem of surveillance is in the case of a very precise kind of thinking, let us now see how the surveillance of surveillance comes to be established.

In its simplest form, intellectual surveillance consists of waiting for a *definite* fact and looking for a *specific* event. Surveillance does not take place with respect to just anything at all. It is directed toward an object that has been fairly accurately indicated, and which at the very least gains from simply being indicated. There is nothing new for the subject exercising surveillance. The phenomenology of newness in the object cannot eliminate the phenomenology of *surprise* in the subject. Surveillance is therefore consciousness of a subject that has an object, a consciousness that is so clear that subject and object define each other, and grow more and more closely united the more accurately the rationalism of the subject prepares the technique for the surveillance of the object under examination. Our consciousness of waiting for a clearly defined event must be accompanied dialectically by consciousness of our mind's freedom, so that the surveillance of a clearly specified event is in fact a kind of rhythmanalysis of both central and peripheral attention. However alert and vigilant it may be, simple surveillance is essentially an attitude of empiricist thought. From this point of view, a fact is a fact and nothing but a fact. The acquisition of knowledge respects the contingency of facts.

The function of the surveillance of surveillance can come into being only after a "discourse on method," when conduct and thought have found their methods and have also given value to those methods. Respect for such a highly valued method will then enjoin attitudes of surveillance that a very special kind of surveillance must then maintain. Surveillance that

Applied Rationalism

is kept under surveillance in this way is at one and the same time consciousness of form and consciousness of information. With this "doublet," applied rationalism makes its appearance. What is required is, in effect, that we should apprehend *formed facts*, facts which actualize *principles of information*.

Moreover, we would note at this point the abundance of material that the teaching of scientific thinking provides for an exponential psychology. The intellectual training of a scientist has much to gain from making surveillance of surveillance quite explicit, for this in fact is consciousness of the rigorous application of a method. Here, a carefully chosen method will play the part of a correctly psychoanalyzed *superego*, in the sense that errors will appear in an atmosphere of serenity; they are a source not of suffering but rather of instruction. We have to commit such errors before the surveillance of surveillance can be alerted and can learn. The psychoanalysis of objective knowledge and of rational knowledge takes place at this level, where it casts light on the relationship between theory and experiment, form and matter, the rigorous and the approximate, the certain and the probable, all of which are dialectics that require *special kinds of censorship* to prevent us from passing from one term to the other without due precaution. This will offer many an opportunity to clear away philosophical blockages; indeed, many are the philosophies that come onto the scene with the express intention of imposing a superego upon scientific culture. If we boast of our realism, our positivism, or our rationalism, we shall, now and then, rid ourselves of the censorship that ought to guarantee the frontiers of the rational and the experimental, and the relations between them. Were we to make constant reference to a philosophy as if to an absolute, we should be bringing a particular censorship into force, the legitimacy of which has not always been considered. The surveillance of surveillance is at work on the boundaries of both empiricism and rationalism, and it is consequently in many respects the mutual psychoanalysis of these two philosophies. The censorship exercised by rationalism and that exercised by scientific experiment are correlative.

In what circumstances might we see the appearance of $(\text{surveillance})^3$? Quite clearly, this would be when it is not just the application of a method that is under surveillance but rather the method itself. $(\text{Surveillance})^3$ requires that *method* be put to the test; it requires too that rational certainties be exposed to all the hazards of experiment, or that some kind of crisis should arise in the way we interpret the phenomena we have duly noted.

The active superego is deeply critical where both these areas are concerned. It is not just the cultural ego that it indicts but also the previous forms of the cultural superego; first, of course, this criticism bears upon the culture imparted by traditional education, and second, it concerns rational culture and the very history of the rationalization of knowledge. To put it in a more condensed form, we can say that the activity of (surveillance)³ enjoys complete freedom with respect to the historicity of culture. The history of scientific thought is ceasing to be a necessary avenue of approach, and it is now just a kind of gymnastics for beginners, providing us with examples of intellectual development. Even when it seems to follow in the wake of a particular historical development, the kind of culture which we have in mind here, that is to say a culture which we constantly hold under surveillance, will recreate recurrently a well-ordered history that bears absolutely no relation to history as it really happened. Here, in this remade history, everything is a value. The (superego)³ can find much speedier condensations than the examples that are diluted in historical time. It thinks history, knowing full well that it would be a very great weakness to relive it.

Should we also perhaps draw attention to the fact that (surveillance)³ has grasped the relations between form and purpose, that it destroys the absolute character of method, and that it judges a particular method as being simply a moment in the progress of method in general? Where (surveillance)³ is concerned, there can no longer be any fragmentary pragmatism. Method must prove that it has a rational finality which has nothing at all to do with some momentary utility. We should at least envisage some kind of supernaturalizing pragmatism, one specifically intended to be an anagogical spiritual exercise, a pragmatism that would look for opportunities to go beyond, to transcend, and that would raise the question as to whether the rules of reason are not in themselves a form of censorship which is to be infringed.

Here we sense that the way is being prepared for a (surveillance)⁴ which would save us from an unreasoned loyalty to admittedly rational ends. Yet this attitude is obviously rare and fleeting. We are simply noting it as a possibility of which we have scant proof. Indeed, the psychology of the scientific mind cannot, in our opinion, provide us with the outlines of this (surveillance)⁴. While the first three exponents of surveillance are, in our view, attitudes of the scientific mind that are relatively easy to observe, (surveillance)⁴ seems to us to approach a danger area. It is in poetry or in a very special kind of philosophical meditation that we should be more *likely to discover the extreme lucidity of (surveillance)*[4]. *Such lucidity is to be found where time is full of lacunae, when the thinking being is suddenly surprised that he is thinking. At moments like this, we really have the impression that there is no longer anything that rises from the depths, nor anything impulsive, or determined by some destiny that springs from our origins. It would seem that we ought to turn toward a philosophy of nascency. And when we allow poets to lead us, we feel the need to establish a fifth element, a luminous, ethereal element, which would be the dialectical element of the four substances which for ten years we systematically dreamed. Yet this desire to fuse together in some way books undertaken in very different horizons betrays, of course, too great a fondness for all that is systematic; but this will surely be forgiven in a philosopher whose rule has always been, though often at considerable cost to himself, absolute philosophical sincerity.*

Le Rationalisme appliqué (1949; Paris: Presses Universitaires de France, 1970), 66–68, 77–81.

"The rationalist is not an angler," Bachelard declares in 1950, addressing a gathering of philosophers, and goes on to say that "you can be very intelligent, yet not be rationalist" (*Eng. rat.*, 59–60). Reason is not innate; it requires work, and it is learned in a "culture." Its discoveries are also prepared in this culture; they do not just "turn up." "This applied, working, socially coherent rationalism confers as a result an extraordinary human value" (58). Here, as always, Bachelard's humanism is indisputable. Indeed, his affirmation of the "human value" of science seems to grow more urgent in his last three books on epistemology, as for instance when he begins *Le Matérialisme rationnel* by considering man in modern science and the "psychological richness" it brings him (1–4). There are no doubt several explanations for this—awareness of a widespread fear of science after the war, of an equally perilous faith in science not just as useful but as a source of power, and possibly too his own advancing years contribute to his sense of the need to speak out more urgently—but, as these extracts have shown, the principal explanation is humanism itself. Bachelard's humanism is polemical, arguing against the humanism of Descartes, Husserl, and Sartre, skillfully turning against them their own weapons. Modern science undermines their view of both reason and reality, and of the human subject. However, modern science does not refute humanism—

it is, as Bachelard constantly demonstrates, man-made—but leads instead to what I have called a subversive humanism. The progress of modern science, its increasing difficulty, and its exploration of more and more "regions of rationality" not only prove the congruence of reason and reality in applied rationalism but also show that this rectified reality and these "objects of thought" change the human subject. This subversive humanism is central to all Bachelard's work, passionately argued in his polemics with other thinkers, while at the same time to him utterly self-evident. To quote another conference paper given in 1950, "it is of course the scientist who makes science, but it is also science that makes the scientist, it is science that educates the scientist. Which came first, the chicken or the egg? Only a naïve philosophy of life could ask a question like that" (*Eng. rat.*, 44).

Chapter 9

Unfixing the Subject: 1957–1961

> The candle's flame is an hourglass that runs upward.
> —*La Flamme d'une chandelle*

In 1961, a year before his death, Bachelard's last book, *La Flamme d'une chandelle*, is published. He begins by describing it as "just a little book of reveries, with no burden of scholarship" (1), setting it apart from his other books on poetry, in particular its immediate predecessors *La Poétique de l'espace* (1957) and *La Poétique de la rêverie* (1960), with their explicit aim of presenting a "phenomenology of imagination" (*PE* 2, *PR* 7). Yet this modest intention is deceptive. *La Flamme d'une chandelle* is indeed a quiet book, but its gentleness, its mellowness of tone rests on uncompromising values. These reveries are not simply the daydreams of an old philosopher: as this extract from chapter 5, "Lamplight," will show, they explore the relationship between men and things, between subjects and objects, and more than this, they are meant by Bachelard to help this to be the right relationship, a "fellowship," a "friendship" as he puts it here, which somehow makes us more fully human.

An oil lamp is "more human" than today's electric table lamp because it gives light "thanks to man's ingenuity" (16); it depends on the care and attention of an individual, without which there is no light. As Bachelard describes in this extract, it will behave differently not just for different people but also at different times for the same person. An electric table lamp, on the other hand, is not affected by the person who switches it on; its light is a function of the structure of the bulb and the amperes of the electric current. It is a uniform object used by a uniform subject. What troubles Bachelard most about this kind of relationship between subject and object is that it impoverishes the subject, making him "the mechanical subject of a mechanical action." Paradoxically, the subject who uses an object, for whom the object is a uniform instrument or tool, is not as a result of this more dominant, more fully a subject, but rather is deprived of his subjectivity. Modern life makes this kind of relationship inevitable: it is convenient to obtain light by touching a switch, and the fact that we play no real part in the production of light is not in the least disturbing. But there is a risk that this convenient passivity will completely overtake our daily lives, that we shall be, as a result, dehumanized. We cannot turn our backs on modern life, and Bachelard is not advocating this. We can, however, break with it, with the help of poetry. Writers such as Henri Bosco, for example, to whom *La Flamme d'une chandelle* is dedicated, restore the reader to a world of subjects, both human and nonhuman, to what Bachelard describes as a world of creatures that create.

This notion of a break with everyday life recurs in *La Flamme d'une chandelle*. Poetry is not merely nostalgia, a harking back to a past which is remembered as more human. Instead, it is something entirely new, breaking with past and present equally, exploring possibility through written language. Bachelard formulates in this book the complex idea of the "image-thought-phrase" (23), the "image-phrase" (72), which disrupts the conventional idea of the image as visual, as belonging to lived experience, whether present or past. More clearly and simply than before, he shows that the poet's images are linguistic. He gives as an example of an "image-thought-phrase" Joubert's "a flame is a damp fire." This juxtaposition of an adjective and a noun which ordinary experience holds as contradictions breaks with "the banality of judgments with regard to familiar phenomena" (23). This "prowess of language" (23) takes us beyond everyday reality to "a human reality" (24). This does not, however, imply the poet-subject's dominance. His language is, so to speak, triggered by the object, by properties of the object apprehended as possibilities: we usually think of a flame as hot, as drying, but put a glass jar over a lighted candle, and the beads of moisture forming on the glass show that it is indeed "a damp fire." The poet therefore restores us to the object, and at the same time to the subject, for in freeing the object from the limits of usefulness, the subject is also freed. Too often, as Bachelard puts it in this passage, we are "caught and crystallized" by adult life; our identity is fixed once and for all as users of objects, consumers who are indeed a strange kind of subject, who, while sovereign in that kingdom of the useful, are in fact entirely without power, subjects fixed in the uniformity and anonymity imposed by our daily lives.

Poetry can restore to us our sense of ourselves as subjects, not as sovereign—this after all is the empty lesson of everyday life—but as Bachelard puts it here, as "creatures."

Bosco shows us that the lamp is more than "just something, an object to be used," that it is a creature, in the sense that it "creates light." Bachelard changes the sense of this word; he inflects it, in the geometrical rather than the grammatical sense, bending its curve from the concave "that which is created" to the convex "that which creates." He exploits the possible ambiguity of the word, the propinquity in French of two words *créateur* and *créature,* the tripping of the tongue and of the eye which makes the one slide into the other. This, then, is a good example of the way Bachelard reads poetry, of how he writes what he reads. His use of language is very like the poet's; he succeeds in loosening words from the bonds of arbitrary definition, so breaking our habitual notions of subject and object, of men and things, as creators and created. In the images of poetry, he writes in *La Poétique de l'espace,* "the duality of subject and object is iridescent and shimmering, it is in active and endless inversion" (4): the subject creates the object, and is in turn created by it; the object is created by the subject, which in its turn it creates. Hence Bachelard's conception here of the subject as creature.

Extract
Lamplight

Whenever we live close to familiar, everyday things, we begin once again to live slowly, thanks to their fellowship, and so yield to dreams which have a past, yet in which there is always something fresh and new. The objects we store away in our treasure chest of things,[1] in our small personal museum of beloved things, are all of them talismans for our dreams. We have only to evoke a much-loved object and its very name will set us dreaming of some ancient story. So, when an old word comes to change its object and name something very different from the old, beloved thing we treasure, our dreams fall to ruin. People who have lived part of their lives in the last century will say the word lamp with very different feelings

1. Bachelard's word here is "chosier": simple, vivid, and untranslatable!

from those of today's generation. *I love to dream over words, and the word "bulb" strikes me as laughable. No bulb can ever be sufficiently familiar for us to give it the possessive adjective.² Who nowadays can say "my electric light bulb" as in bygone years he would have said "my lamp"? How are we to go on dreaming when there is such a decline in possessive adjectives, in the adjectives that spoke so clearly of the fellowship we enjoyed with our own objects?*

The electric light bulb will never let us dream as once we did with the lamp before us, the living lamp that took oil and made it into light. We are now living in the age of carefully dispensed light, in which the only part we play is to touch a switch. We are no more than the mechanical subject of a mechanical action. We can no longer benefit from this action and make ourselves, with rightful pride, the subject of the verb to light.

In that beautiful book of his Vers une cosmologie, *Eugène Minkowski has a chapter entitled "I Light the Lamp."³ Yet here the lamp is an electric light bulb. A finger on the switch is all that is needed for darkness to be immediately replaced by light. The very same mechanical action will result in a change in the opposite direction. A little click says both* yes *and* no *in exactly the same tone of voice. Thus, the phenomenologist has the means whereby he can place us successively in two worlds, and therefore in two consciousnesses. We can play endless games of* yes *and* no *with an electric light switch. However, by accepting something mechanical, the phenomenologist loses the phenomenological richness of his action. These two realms of darkness and light are separated by a single instant only, an instant which has no reality, a Bergsonian instant, an instant conceived by an intellectual. This instant was far more dramatic when lamps were more human. When the old lamp was lit, we were always in fear of our clumsiness or our ill luck. The wick is never quite the same as it was the previous evening. It will begin to smoulder if we neglect a single detail, and if the glass is not placed absolutely straight, the lamp will begin to smoke. We always have much to gain by granting familiar things the friendship and attentiveness they so well deserve.*

2. Bachelard's footnote: Jean de Boschère is very sarcastic about a scene in which an "electric light bulb" venerates the Virgin's statue, instead of a little lamp. A lamp is surely a gaze: "a lamp should burn in the dark eyes of its oil" (cf. *Marthe et l'enragé* [Paris: Emile-Paul Frères, 1927], 221). An electric light bulb cannot gaze. Note amended.
3. *Vers une cosmologie: fragments philosophiques* (Paris: Aubier, 1936).

Unfixing the Subject

It is through the friendship that poets show toward things, toward their own things, that we may come to know those clusters of instants which give ephemeral acts a human value. So Henri Bosco shares with us his childhood memories, and in their telling, the lamp regains all its former dignity. The lamp keeps faith with us in our solitude, and Bosco writes of it like this:

> We soon realize, and not without emotion, that *the lamp is some one*. By day, we thought it was just something, an object to be used. But when daylight fades and we find ourselves in a lonely house where gathering shadows mean that we must feel our way along the wall if we are to move, then we go in search of the lamp we cannot find, only to discover it in a place we had forgotten; the lamp we have found at last and that our hands now closely grasp offers us its gentle, reassuring presence, even before it is lit. It brings us peace of mind, it thinks of us . . .[4]

A description such as this will be quite meaningless for those phenomenologists who define the being of objects in terms of their "utensility." They have coined this barbarism in order to end once and for all the seductiveness of objects for us. Utensility is, in their view, something we can know so precisely that it has no need for this kind of dreaming over memories. Yet memories deepen the fellowship that we enjoy with kindly objects, with objects that keep faith. Every evening, at the appointed time, the lamp does its "good deed" for us. These emotional inversions between a good object and a good dreamer will be readily condemned by a psychologist whom adult life has caught and crystallized. In his opinion, these inversions belong quite simply to an infantile stage in life. Yet as the poet writes, their poetic sense stirs once again and comes to life. The writer knows that his words will be read by spirits sensitive to the first poetic realities. Bosco continues his description of the lamp like this:

> Watch the lamp closely when you light it and tell me if, secretly, it is not the lamp that lights itself before our inattentive eyes. I might perhaps cause some surprise were I to declare that the lamp does not so much receive the fire we bring it as offer us its own flame. Fire here is something quite external. It is simply an opportunity, a convenient excuse of which the

---

4. Bachelard's footnote: Henri Bosco, *Un Oubli moins profond* (Paris: Gallimard, 1961), 316. My translation; this book has not been translated.

closed lamp takes advantage so that it may give of its light. The lamp is. I experience the lamp as a creature.

The word "creature" is decisive here. The dreamer knows that this creature creates light. It is a creating creature. We have only to give it one good quality, we have only to remember that it is a good lamp and our lamp will come alive. It lives in the memory of that peace we once knew. The dreamer remembers the good lamp that would light so easily. Here, the verb "light" shows even more plainly that the creature giving us light is to be valued as a subject. Words, with their tender inflections, help us to dream well. To all things, give their qualities, and to all active beings give, freely and unstintingly, the gift of their true and rightful power, then will the universe be glorious. Take a good lamp, a good wick, good oil and we shall have a light that makes glad the heart of man. He who loves a fine flame also loves good oil. He follows where he is led by all those cosmogonic dreams in which each and every object in the world is the germ of another world. For someone like Novalis, oil is the very substance of light, the rich yellow oil is condensed light, condensed light that seeks to expand. Man comes, and with a frail little flame he liberates the forces of light that are locked in the prison house of matter.

Our dreams will probably no longer go so far as dreams like this. Yet we have had such dreams. We have dreamed of the lamp that gives light and life to some dark substance. How, too, could someone who dreams over words fail to be moved when he learns from etymology that petroleum is petrified oil? From the very depths of the earth, the lamp brings forth light. The older the substance upon which it acts, the more sure we may be that the lamp is being dreamed in its true character, as a creature that creates.

La Flamme d'une chandelle (1961; Paris: Presses Universitaires de France, 1964), 89–94.

Poetry can restore to us this sense of ourselves as "creatures," as subjects beyond the conventional limits of subject and object. Images of candles and lamps, for instance, show these simple, familiar objects as possessing "an abundant multiplicity" (12) which Bachelard puts before us in *La Flamme d'une chandelle*. This has nothing to do with multiplicity of use, nor is it simply a trick of what he calls "prolix comparison" (71). It is the work of

Unfixing the Subject

the *written reverie* of both poet and reader, which is not prolix but peculiarly terse. What he calls the "adventures" and "exploits" of reverie (12, 39) do not take us into "faery worlds forlorn." They are first and foremost "bold" language (75), "polemical actions against common sense, slumbering in its habits of seeing and speaking" (72). Poetry is language which has broken with immediate experience and which, because of its particular "flexibility," can through "syntheses" and "fusions of objects" (76) reveal their "abundant multiplicity." Even more striking than Bachelard's insistence on poetry as language is his declaration that these linguistic images of candle flames "nourish" the reader (4, 56). This new metaphor implies rather more than the by now familiar idea that the subject is changed by the object: it raises the question of how this change takes place, of the instrument, the means of change. If the candle flame "nourishes our verticality" (4), this is not the consequence of either perception, intellection, or even imagination, and certainly not of some unconscious communication. This "nourishing" takes place through language, because the candle flame is a linguistic object, an object that is in, through, and against language, that is language and nothing more. It is therefore language that "nourishes" the reader's verticality, but how? Above all, it complicates and enriches a dimension that appears simple and sometimes cruelly stark; it can hold together contradictions which threaten to destroy our verticality, Bachelard's metaphor for our progress, for our human being. The candle flame for the poet is not just a useful source of light: it gives light to others but itself is solitary (chap. 2); it creates light and is destroyed as it creates (24–26); it flickers, it is fragile, yet remains vertical (58). Through language, the candle is taken by the poet, by the "dreamer," beyond its own reality to "human reality" (24). It expresses "human values"; it "nourishes" the reader's verticality by teaching him to live against, to live beyond life.

This desire to subvert lived experience also pervades both *La Poétique de l'espace* and *La Poétique de la rêverie*. Bachelard again rejects the psychoanalytical approach to reading because of the importance it gives to writers' past lives. Phenomenology now seems more congenial, more in keeping with his wish to attend to the facts of reading, to the reader's consciousness of written images. Yet since phenomenology is itself a philosophy of lived experience, Bachelard's adoption of it may well seem inconsistent. He already made explicit use of phenomenological concepts in his last three books on science, and this should be borne in mind here. Husserlian intentionality in particular seemed to Bachelard inadequate to the "rational consciousness," and therefore he modified it. He also modified Husserl's *epoche*. This was the first phenomenological concept to which he made explicit reference: antedating *Le Rationalisme appliqué*, it suggests why he was drawn to Husserl's ideas. He uses it first in *La Philosophie du non* in 1940: "contemporary thinking begins with an *epoche*, with a bracketing of reality" (34). At this stage, Bachelard understands this "bracketing" as a break with immediate reality, and it is clearly this that attracts him, although for Husserl in *Les Méditations cartésiennes*, for instance, the world—immediate reality—is still there; it is for me, part of my lived experience (Meditation I, section 8). Bachelard in due course understands this; he subverts and opens up the *epoche*, making it a complete break with life, with the past, making consciousness not just rational but "instrumental," consciousness of creating through scientific reason a reality that is neither lived nor "already there" but new (*ARPC* 6–11). This, then, is the background to Bachelard's phenomenology of imagination, which can be described as a non-Husserlian phenomenology that is quite explicitly not a philosophy of lived experience.

Given all this, how are we to understand Bachelard's description of reading poetry as "living the unlived and opening oneself to an openness of language" (*PE* 13)? We must take care not to interpret "living" too hastily. It is perfectly clear that Bachelard is not a partisan of lived experience. If he uses the word here, it is to make two points: first, that reading is an *activity;* second, that this activity obliges us to redefine what we mean by "life." The "unlived" of poetry is not a fantasy world; living the unlived is not escapism. The unlived is language, and it reminds us that life involves more than perceiving, thinking, using, that we live in what Bachelard calls a *"linguistic space"* (*PE* 11). "Everything that is specifically human in man is *logos,"* he writes; "we can never manage to meditate in an area that exists before language" (*PE* 7). Life conceals this from us, making words into tools or labels, giving things precedence over words, enclosing words in meaning (*PE* 138). But words, Bachelard says, are like houses, "with cellars and attics," where "common sense lives on the ground floor," abstraction in the attic, reverie in the cellar. Between cellar and attic is a staircase, and this, the staircase of language, is the poet's space, "the poet's life" (*PE* 139). In *La Poétique de la rêverie*, he describes the "true poet" as "bilingual," speaking "the language of meaning" and "the language of poetry" (160). What, though, is the language of poetry? In both languages, the words are the same, but the poet frees words from meaning, he *opens* words, he

Unfixing the Subject

exploits the "dialectic of the open and the closed that is in language itself" (*PE* 199), he unfixes language from meaning, from referents, and restores its autonomy. Bachelard himself engagingly describes his own discovery of the autonomy of language like this: "I am a dreamer of words, a dreamer of written words. I think I am reading. A word stops me . . . it gives up its meaning . . . it takes on other meanings . . . words go off into the thickets of vocabulary, looking for new companions, getting into bad company . . ." It is even worse, he says, when he writes (*PR* 15–16)! This autonomy is not the whole of language. Bachelard does not deny the claims of meaning, the exigencies of social and practical life, but he reminds us of the dialectic of meaning and possibility within each word, of what he, in his idiosyncratic way, calls the "staircase of language."

Bachelard has believed from the beginning that "the chief function of poetry is to transform us" (*L* 105), trying in different ways to explain how this transformation takes place. He has always regarded it as being largely the work of language, but it is in his last books that he commits himself unequivocally to the conception of human beings as *logos*. This does not lead him to decenter the subject; the autonomy of language revealed by poetry is not threatening but enhancing. If words play between meaning and possibility, then we ourselves—the dwellers in "linguistic space"—must do so, we too are a dialectic of closure and openness. Hence, very simply, Bachelard's conception of man as a "half-open being" (*PE* 200), "on the surface which separates the region of the same from the region of the other" (*PE* 199), as an "unfixed being," a "spiral" (*PE* 193).

Chapter 10
Gaston Bachelard: Subversive Humanist

> The new thought and the new poetry require a rupture and a conversion. Life must wish for thought. No value is specifically human if it is not the result of renunciation and conversion.
> — *Lautréamont*

Bachelard's polemical way of thinking has struck a number of his commentators as his most constant characteristic.[1] He is also thoroughly subversive, which is rather different and, as I have tried to show, more fundamental. Rather than simply argue, Bachelard takes ideas from other thinkers and uses them in such a way that they undermine their original sense. An example is his treatment of Bergson's "creative evolution." This is not dismissed, even though he is in total disagreement with Bergson here; but instead, he subverts it, for by making creative evolution the work of an "intellectual" rather than a "life force," he both undermines and *opens* Bergson's thinking. His long polemic with Bergson is marked by this subversiveness, perhaps at its most radical when time and consciousness are conceived as discontinuous. There are many examples of this in Bachelard's work; indeed his last three books on science show its energy and subtlety undiminished as he takes on philosophers of existence in general and more specifically Husserl's phenomenology. What first attracts Bachelard to Husserl is, we know, the idea of the *epoche;* later it is his conception of consciousness, and more generally Husserl's view that philosophy should be guided by "an authentic science" (*Méditations cartésiennes*, I, section 3). However, he regards as invalid and untenable Husserl's idea of "true science" as "universal" and "unchanging," as having an "absolute foundation" (I, section 5). He does not, though, refute Husserl, for refutation is not the aim of his polemic. Instead, Bachelard shows what happens when science is allowed to lead philosophy: the old idea of universality has to be revised in the light of "regional rationalism," and the value traditionally attributed by philosophers to first principles, to founding knowledge, must be rethought, given the progress and process of modern scientific knowledge. Bachelard therefore subverts Husserl by showing that science is a "constructive activity" (*Eng. rat.*, 65) where truth is in "summits," not foundations, and following on from this, by his conception of the rational consciousness where subject and object are interdependent. Where Sartre is concerned, Bachelard's hostility is unmistakable in his last three books on epistemology. Yet once again he does not simply oppose him. He reinterprets the ideas of situation, commitment, and freedom, using them against Sartre: if we are situated in and committed to modern science, if freedom is to think, then lived experience emerges as something that endangers "human reality."

The philosophers Bachelard subverts in this way can all be described as humanists, so that paradoxically he is both the opponent and proponent of humanism. His attitude to philosophy and philosophers in general is crucial here. He is often scathing about what is after all his own profession, about "intellectual Robinson Crusoes" (*Geneva* 26) and "the Knights of the *Tabula Rasa*" (*Geneva* 28), deeply critical of the French philosopher's traditional concern with first principles, with unity and universality, identity and permanence, and with individuality (*Eng. rat.*, 41–44, 49–50), critical too of contemporary philosophy for its "intoxication with personality, with originality" (*Eng. rat.*, 36). However, these criticisms do not lead him to reject philosophy; rather he seeks to redirect it. He comments on the lack of interest among his contemporaries in the philosophy of science, an indifference which he explains by the general feeling that science is "tainted by utilitarianism" (*Eng. rat.*, 35). These same philosophers continue to talk of reason and reality, forgetting that these ideas are affected by science. Bachelard's whole enterprise is to demonstrate that contemporary science invalidates what he calls "philosophical rationalism" and "philosophical materialism," that if philosophy is to be worth anything, it must pay attention to what is happening in science. In his view, science transforms traditional and contemporary humanism. Bachelard constantly stresses the role of reason, of mathematics, in modern science, and the creativity of human beings, but he also makes it clear that scientific knowledge is not grounded in human beings. He describes the modern scientist very simply as a "patient weaver," working at the "loom of science," where he combines the "warp of reason" and the "weft of experiment" (*Eng. rat.*, 43).

This last quotation pinpoints another aspect of Bachelard's subversiveness, namely his use of language. He delights in words, he has a great gift for words, and some might feel this linguistic creativity to be rather flashy self-indulgence. This, though, would be out of character for Bachelard. Jean-Claude Margolin has described the "scandal" of Bachelard's language for his contemporaries, for those traditional philosophers who liked to remain in their own systems, safe in their own vocabulary, who objected to Bachelard's use of, for instance, psychological, psychoanalytical, or sociological terminology in the philosophy of science (1974: 28). Bachelard uses words to shock his readers into thinking differently, to unfix them from the habits and systems he regards as harmful. But words do inevitably belong to systems; their definition and use are bound up with habit. "Reason," for example, already has meaning before Bachelard uses it, so that although he can qualify it in various ways, most often he simply has to use the word as it stands, his own interpretation being first made clear by the context. The context is usually dense, difficult argument; it is also often extensive, more a perspective of several books than just a context properly speaking, and this may well lead to misunderstandings if his books are read quickly or in isolation from each other. This would appear to have happened in the case of Marxist critics of Bachelard who, perhaps because of their own convictions, are unable to understand how he is using words like "reason" and "matter." It is a problem inherent in language, for which one can hardly blame Bachelard, but he does not make things any easier by refusing to be "labeled." Labels, he says, refer to past preoccupations; he does not wish to be labeled—in this case as a psychoanalyst of reason—because, as he puts it, "I am reorganizing myself" (*Eng. rat.*, 50).

This inconvenient and disconcerting refusal to take sides, to wear the badge of some "-ism," is perhaps the ultimate subversion where French philosophy is concerned. Vincent Descombes has drawn attention to the importance of "opinion" in the reception of recent ideas in France, to the politicization of philosophy in a country where its practitioners are in the main teachers, and therefore state-employed (1980: 5–8). Bachelard does not fit into this scheme of things. He was a university teacher, he was a philosopher, but he was not political, that is to say he did not follow a particular party line, refusing in general to be partisan. His lack of politicization is plainly worrying to some of his fellow countrymen. "Bachelard,"

Michel Vadée declares in the first sentence of his introduction, "has become a problem for Marxist philosophers in France" (1975: 11), going on to conclude, after what is often a perceptive and illuminating discussion, that Bachelard is a "bourgeois rationalist," "the representative of a conservative, retrograde ideology" (275). Almost automatically, it would seem, Bachelard's rationalism is politically suspect. This kind of Marxist reflex may well explain why Bachelard did not involve himself in party politics, for this would have meant fitting into a system and being fixed by it.

Yet Bachelard did belong to the academic system in France. Professor at the Sorbonne, member of the Institut de France, to which he was elected in 1955 as one of the forty members of the Academy of Moral and Political Sciences (one of the Institut's five constituent bodies), with his doctorate, *agrégation,* and *licence,* he had been a teacher since 1919, when he began his career at the *collège* in Bar-sur-Aube. Bachelard's participation in this particular system came fairly late, however, for he was thirty-five when he started schoolteaching. Perhaps more important, his upbringing was not that of the *bête à concours* (competitive examination fodder); he was not trained from an early age to make academic success a be-all and end-all. This background, together with the fact that he was largely self-taught, detached him from the academic system and its politics. He obtained his mathematics degree in 1912 while working in the post office in the Gare de l'Est in Paris (1907–13), his *licence* and *agrégation* in philosophy (1920 and 1922) and also his doctorate (1927) being gained while a school science teacher in the provinces (1919–27). The fact that Bachelard continued his studies to a high level in such circumstances shows that he was highly motivated. His lack of political involvement was a matter not of apathy or indifference, but rather of being already committed to something else, that exceeded the limits of politics, that is to say to knowledge, to the life of the mind. He is quite explicit about this, affirming that "my act of scientific faith and all I aspire to can be expressed like this: intellectual destiny is a *destiny;* understanding is not a means, it is an end" (*Ghent* 4). Bachelard firmly believes that man is an intellectual, not a political animal, that intellectual needs far outweigh social demands. This is reflected when he says—twice—that "society was made for the schoolroom," and not vice versa, that school was an end, indeed *the* end (*Geneva* 29, *FES* 252). School ought to be seen as preparing children not for life, but for knowledge (*Geneva* 29). One may well ask what kind of knowledge he means here. A consideration of Bachelard's apparently ambiguous attitude to teaching will help us to understand his values. On the one hand, he defines himself as "more of a teacher than a philosopher" (*RA* 12), his last lecture at the Sorbonne in 1955 ending with the words "I do not leave the Sorbonne with a merry heart. I have given myself to teaching" (Lescure 1963:130).[2] On the other, he is often scathing about teachers, about traditional academic values in France. Reason and clarity were greatly esteemed, but what was taught was an established, formed reason, and the clarity of past ideas. Teaching is therefore usually regarded as handing on and repeating an intellectual heritage, the effect of this being for Bachelard to close pupils' minds (*Ghent* 13, *Eng. rat.*, 40). In his view, traditional education prevents young people from reaching their potential, and traditional academic values, especially that of rationality, are obstacles to the fulfillment of their "intellectual destiny."

Bachelard is therefore apolitical yet committed, his commitment being not social but intellectual. He is committed to a particular kind of teaching, which involves the subversion of values which are still regarded as the hallmark of educated, cultivated minds: rationality, clarity, coherence, consistency. In *L'Intuition de l'instant*, his aim is "a pedagogy of discontinuity" (56), and in *Le Nouvel Esprit scientifique*, "a pedagogy of ambiguity" (19), based on the "lessons" of wave mechanics (88, 97), and of the "new scientific mind" generally. He is equally subversive where traditional literary education is concerned, describing the teacher of literature in *Lautréamont* as "*correcting man,*" whose aim is to control, censure, and limit language in the name of clarity (64). This taught language closes children to language itself and also to the experience of literature. As an antidote to this, he recommends what he calls "a pedagogy of the imagination" (*L* 155). Indeed, this "pedagogy of the imagination" is the aim of Bachelard's books on poetry, in which he seeks not to inform but to transform the reader. He regards reading poetry as "a kind of physical hygiene," as "gymnastics for the imagination" (*ER* 249), phrases remarkably reminiscent of his description of the effect on him of reading Heisenberg as "excellent mental hygiene," the result of having to grasp the paradoxes of wave mechanics and the dialectical relationship of matter and energy (*NES* 88). Reading science and reading poetry are both ways of overcoming the "ankylosis" of everyday life and thought (*NES* 43, *PN* 104), its "psychological hardening" (*NES* 87, *AS* 278). This sort of vocabulary makes Bachelard's values very clear: we are impaired by lived experience and, because they are rooted here, by the values we are taught. His ambition is to help us to "mobilize" reason and imagination, to unfix us and keep us unfixed.

This bringing together of science and poetry has worried many who have written about Bachelard, causing a general division into two camps, one arguing for and one against unity in his work. Bachelard himself, especially in his later books, tends to make contradictory statements, insisting that imagination and reason are opposites (*PR* 46–47), while having already undermined that statement some forty pages earlier by referring to his "more general philosophical thesis" which is, he says, that consciousness changes and recreates the conscious being (5). Yet is he in fact contradictory and inconsistent? Again, the attitude of his commentators is instructive: unity is regarded as a virtue, difference as a vice. If Bachelard himself stresses in his last books on poetry the *difference* between the two aspects of his work, that there is a break between them (*PE* 1), this is perhaps out of a desire to maintain that difference which he has always regarded as essential to human being. This wish to cultivate difference, so fundamental in Bachelard's thinking, is once again deeply subversive of our habits as readers, trained as we are to seek unity and value consistency. Earlier in his work, Bachelard was fond of comparing science and poetry. He did so in order to draw attention to the creativity of modern scientific thinking, and once again as a kind of shock tactics to jolt his readers out of a positivistic and utilitarian view of science. His references in *Le Nouvel Esprit scientifique* to "the poetic strivings of mathematicians" (35) and to Mallarmé's images (60) were no doubt bad enough for hardened rationalists, but it must have been odder still to the ears of philosophers attending a conference in Prague in 1934 when he compared chemistry—which he himself had described as "that most experimental and positive of sciences" (*PCCM* 7)—to poetry, declaring that "some chemical bodies created by man are no more real than the *Aeneid* or the *Divine Comedy*" (*Ét.*, 83). Even more wayward is his reference in *La Philosophie du non* to "mathematical reverie" (13) and his description of Dirac's work on negative mass as a kind of "reverie" (38–39). In a similar way, Bachelard introduces scientific language into his books on poetry, and makes equally unexpected references to mathematics. In *Lautréamont*, for example, he suggests that some poems should be understood "as independent systems just as non-Euclidean geometry is understood as having its own axiomatics" (97), a disconcerting but illuminating comparison which underlines his conception of poetry as "the antithesis of life" (102).

This transgression of the boundaries of science and poetry is an important aspect of Bachelard's work. However, he is not simply a man of C. P. Snow's "two cultures": to describe him as such would be a serious misrepresentation of what he does. Snow was concerned with the subject matter of science and poetry; he wished for poets and ordinary educated people to be better informed about the facts of science, for twentieth-century art to be inspired by twentieth-century science. Bachelard can be seen as reversing Snow's ambition, for if he compares twentieth-century science to poetry, it is in order that scientists may understand that their activity is as creative, as imaginative, as that of poets. In his books on science, his aim is explicitly pedagogic. It is not though to teach us, for example, the second law of thermodynamics, but instead to show us how the facts of modern science affect our status as human beings. "Contemporary science," he says, "creates a new nature, in man and outside man" (*Eng. rat.*, 99). "Those who enter modern scientific culture come, quite obviously, under the rule of the human" (*Geneva* 21). "Modern science is essentially a humanism" (*Geneva* 26). Bachelard never tires of saying this, and it remains the central, urgent lesson he feels duty-bound to teach.

He is aware that this conception of the "human value" of science is both controversial and open to misunderstanding by his contemporaries, that in particular it is likely to give rise to the charge of idealism. From first to last, he made it clear that reason in modern science is not what he calls the "traditional" reason of philosophers, that it is not a priori, not general and universal, not founded in the human mind. Idealism is a Circe, as he puts it (*Eng. rat.*, 65), but the rationalism of modern science saves all of us from shipwreck: were we all-powerful, we would face no problems, and "reason would cease to breathe" (*Eng. rat.*, 51). Paradoxically, Bachelard's humanism is *against* the conception of the sovereign, founding subject. Paradoxically, too, he is now severely censured by those who also refuse this view of the subject. Our contemporaries seem to be as fixed in their own preconceptions and prejudices as the philosophers Bachelard chides in his own day—and this in spite of the lip service paid nowadays to discontinuity—every bit as obedient to conditioned reflexes, unable to grasp the possibility of a nonfounding reason and a nonfounding subject. Bachelard's Marxist critics have been the most censorious, Vadée regarding him as an "obstacle" to Marxist truth (1975: 7–9), dangerous because he is an ideological wolf in a dialectical materialist sheep's clothing, Lecourt disappointed, thinking to begin with that what he calls Bachelard's "historical epistemology" (1969: passim) is open to a "materialist reading" (1972: 20), but eventually concluding in *Bachelard ou le jour et la nuit* (1974) that despite being so near the truth, he fails to be either scientific or materialist. No doubt Bachelard's insistence on the dialectical relationship of man and matter led them to expect rather better of him. Yet he himself distinguishes between his dialectics and Hegel's, considering the latter as a priori and too general (*Eng. rat.*, 8, *PN* 135–36). Marx, as Henri Gouhier has pointed out, was not a problem for Bachelard and his generation, and the consequent absence of direct discussion, especially in so wide-ranging a philosopher, is tantalizing.[3] However, his discussion of "philosophical materialism" in *Le Matérialisme rationnel*, in particular in chapter 2, "The Paradox of Philosophical Materialism," and chapter 4, "Compound Materialism," would seem to embrace Marxist theory, and Vadée certainly takes this to be so. Bachelard is critical of "philosophical materialism" for ignoring modern scientific materialism; Vadée is equally critical of Bachelard for not placing scientific activity "in contact with a reality that is outside us, that is material and independent of us" (1975: 179). Bachelard ignores nature, he says, and censures him for this. Vadée is plainly not open to argument. Yet science has itself subverted the notion of reality as *either* external *or* internal, and indeed the whole idea of the "natural," as Bachelard shows in relation to contemporary developments in physics in particular. We today are perhaps even more aware that the "rational" and the "real" are no longer distinct, and this not in an abstract, alien area like particle physics but in, for example, genetic engineering which, with its consequences for medicine, agriculture, and pharmaceutics, touches us more closely. Even more important, what is "natural" is increasingly called into question, by developments in the biological sciences especially, not just as a scientific or philosophical problem but as a moral and a legal issue.

Modern science is undeniably difficult, becoming ever more specialized, and because of this, some regard it as excluding, even endangering ordinary people. Colin Smith, for example, concludes his discussion of Bachelard in his *Contemporary French Philosophy* by referring to the dangers of "the derealization characteristic of open rationalism," quoting in support of this a passage from Hannah Arendt's *The Human Condition*, in which science is presented as depriving men of meaningfulness because it cannot be discussed by them but only undergone (1964: 111–12). "Meaningfulness" for Arendt is quite explicitly the province of speech, of talking together about the lived world, with the result that there is a dangerous tension between the truth of science, *beyond* speech, and the truth of man, *in* speech. As we have already seen, Bachelard himself takes a very different view. In his last three books on epistemology especially, he accepts that science is increasingly difficult and specialized, he is aware that for many this undermines the human value of science, but he argues that it in fact confirms rather than denies this value. Specialization *attaches* the subject to an object, he says, by *committing* him to a particular field of study, but this does not lead to fixity, for the progress of science constantly unfixes the subject (*ARPC* 20–23). "The mind has to open up in all directions in order to serve a specialization . . . ," he writes in a contemporary article; "today's specialization requires immense reading and a thirst for new information" (*Eng. rat.*, 41). Habit and received opinion lead us to speak of specialization as narrow and limiting, yet surely we know from our own experience that this is untrue, that as Bachelard puts it, "specialization is not narrow . . . at the very moment when you become specialized, you realize that your mind is opening up" (*Eng. rat.*, 56). To take an example, when, reading Bachelard, we try to understand something very specialized like Heisenberg's uncertainty principle or Pauli's exclusion principle, this opens up a new world of possible knowledge to us. Complete understanding of these concepts is no doubt beyond most of us, but this does not matter to Bachelard. Failure to understand he regards as something positive. When a scientific method fails, he writes, it means that something new has been found, for "failure is a new fact, a new idea" (*Eng. rat.*, 39). In exactly the same way, our failure to understand a difficult idea brings an experience of newness, of openness.

This idea of failure plays an important part in Bachelard's thinking; though it appears simple enough, it can be seen to overlay other ideas and values, to concentrate his subversiveness. It serves in his first book, his *Essai sur la connaissance approchée* (1928), in his arguments against idealism and pragmatism, against in particular the notion of the sovereign subject. Failure to understand is a fact that can be explained only by "rebellious matter," by an "inexhaustible" reality, and so leads him to conceive of knowledge as an "alternating current" between an interdependent reason and reality, two poles held together in a "minimum opposition." Failure also serves in his argument against Bergson, having in Bachelard's view ontological as well as epistemological consequences. Failure to understand brings with it consciousness of our own discontinuity, our own incompleteness, of being bound to something beyond ourselves, to the world, to "newness," without which we have no being. Besides current philosophical systems, Bachelard also questions and seeks to undermine certain values that we are taught but that he believes both ill founded and harmful. Failure is something we have all learned to fear and to judge, to criticize in others and avoid in ourselves. But why? It seems to be in some way an ontological threat, and the self-doubt and at worst self-destruction to which it leads will bear this out. Failure threatens our sense of our own completeness, of the coherence and control which we have learned to regard as the preeminent and defining characteristics of human beings. Traditional humanism cannot cope with human failure. Yet it is a fact of human experience, which Bachelard faces and uses, making it central to his own subversive humanism. Failure to understand means incompleteness, but incompleteness from another point of view is openness, possibility, progress. It is also the discovery of the unknown in both subject and object, a discovery which can be not threatening but invigorating if, and only if, it is an active failure, if it initiates work.

Simple as it is, this idea of work is important in Bachelard's thought, and particularly revealing with regard to the question of his humanism. Something of its implications can be seen when he defines his philosophy of science as an *"open philosophy,"* as "the consciousness of a mind which founds itself as it works on the unknown and seeks in reality contradictions to previous knowledge" (*PN* 9). If there is to be progress in scientific knowledge, there has to be a kind of methodological, constructive failure. This active, working failure can be seen as decentering the thinking subject, yet at the same time sustaining him. It is a decentering the subject initiates and it therefore neutralizes the sense of losing control, but it is, even so, initiated in response to problems posed by "the unknown," by reality, problems which Bachelard regards as "founding" the mind, reconstructing it as it works toward solutions. The work of the mind produces not just new scientific knowledge but a new consciousness. What he calls the "will to coordinated work" is valued for the coherence it brings to the subject, not in the conventional sense of unity, continuity, sameness, but in the "psychological richness" of discontinuity and diversity (*MR* 1–2).

Bachelard indicates at the beginning of *Le Matérialisme rationnel* (1953) —his last book on epistemology—the value the work of the scientific mind has come to have for him, and it is worth quoting from this at length because he places himself quite explicitly *against* traditional humanism, alluding to Montaigne's view of man in the first phrase here, the notion that man is "diverse and changing":

> It is useless to keep saying that man is diverse and changing, for it is a faint, weak kind of changing, and man's contingent diversity does not manage to hide deep poverty. If we are to find true psychological richness in man himself, a sure way is to seek that richness at the summit of thought. There, man can be seen in his will to coordinated work, in the tension of the will to think, in all his efforts to rectify, diversify, and go beyond his own nature. As for tangible proof of this "going beyond," do we not see this in his going beyond ordinary experience, beyond nature itself? Like it or not, everything in man is coupled with knowledge. Knowledge is a plane of being, it is the plane of potential being, a potential that grows and is renewed as knowledge grows. Modern science takes man into a new world. If man thinks science, he is renewed as a thinking being. He accedes to an undeniable hierarchy of thought. His diversity is not simply in the contingent life envisaged by someone like Montaigne. It is "vertical," hierarchical diversity. (1–2)

Scientific thought is valued for its rigor, its constant rectification and progress, its order and coherence, all of which are summed up in the metaphors of verticality and hierarchy. These are by now familiar but still instructive metaphors, exploiting as they do the tensions between our sense of continual, orderly development and of our contingent linearity, subverting the spiritual values that traditionally belong to verticality, hierarchy, and transcendence. This is not though in order to enhance the human at the expense of the divine, but rather to set against the old humanism of lived experience, of the sovereign, identical subject, what Bachelard regards as the new humanism of the "thinking being," thinking and working in and against modern science, and in this way achieving "psychological richness." It might seem that this excludes most of us, the nonscientists. On the contrary, however, Bachelard believes that any one of us can enter science by reading. This in its turn will require "the tension of the will to think," "coordinated work," which is difficult and likely to fail. Failure to understand is unimportant, though; what matters is the work of the mind as it encounters the close pattern of reference points that constitutes scientific argument. Reading modern science is hard but very productive work, which allows us to experience, more than ordinary life ever can, a precise, ordered coherence which because it is constantly being rectified is always reordered, the coherence of ordered possibility and coordinated discontinuity, an open pattern of difference.

But what of reading poetry? Bachelard does not regard this as understanding ideas, nor does he read poems as a whole, as structures of images, so that failure, work, and coherence might seem to play no part here. Bachelard's theory of imagination is not systematic, and his practice of poetry—his reading—appears very free and undisciplined. It is quite incorrect, though, to consider his reading as just "reverie," as giving us all the "right to dream." He is a writer; his books on poetry are carefully composed, both in the sequence of chapters and in the weaving together of many different images within each chapter and on every page. Reading poetry, in Bachelard's case, leads to the work of writing. He would have everyone read poetry "pen in hand," in order to ensure that reading is active, that it is work. He does not, though, insist that the reader of science should write what he reads. Reading science requires a disciplined application of the mind to ordered thought, so that when, however haltingly, we think science, we benefit from its coherence. Poetic images are coherent; Bachelard always stresses this, trying to account for it, especially in his first books on poetry, seeking to help us understand and experience the coherence of apparently incoherent images. He explicitly distinguishes between his approach and that of teachers of literature who look for continuity, setting out himself to show the coherence of very different images, and of discontinuity. Yet though poetic imagination is in its way as coherent as scientific thinking, Bachelard does not discipline his mind to its patterns, and so risks neglecting the very coherence he values. But he does write, and by this active reading, produces a new coherence. Reading poetry like thinking science is a way of constructing coherence, of breaking the circle of lived experience.

This idea of work, of the work of thinking science and of reading poetry, brings us back to Bachelard's subversive humanism. He declares, toward the beginning of his last series of lectures at the Sorbonne in 1953, that the "destiny of humanity can be seen in the perspective of work," using this idea of work quite explicitly to attack the philosophers of being (Lescure 1963:120). Here, work is interpreted as breaking with being, as the construction of possibility. It is a thoroughly subversive idea, which undermines, besides the philosophies of being, pragmatism and Marxism especially, realism and rationalism as well. Human beings can be defined very simply as workers who, as Bachelard puts it, provoke the world by the instrument of reason and imagination (*ARPC* 141), creating and exploring possibilities in the world and in themselves. Their work neither decenters nor centers them because it is against the world and at the same time against themselves, against their own past and their own present. This for Bachelard is the human condition, that work should break with life. His "reader's prayer" is well known: "give us this day our daily need" (*PR* 23). We smile, but recognize that he is serious, for without this need, we shall not seek the sustaining "other," and we shall starve.

Bachelard's last book, *La Flamme d'une chandelle*, ends with a brief self-portrait of himself as worker, remembering his "working past" (107) and with a question about his working future (112). The image he has of himself in his mind's eye is, he says, a *"primary engraving,"* a picture that has meaning for all of us without having to be explained, that touches us because it is to do with our human being: he sees himself alone at his worktable, by lamplight, in a dark room, where there is just one small circle of light (108). Too often, he says, that solitude has led to dreaming and away from "the *adventures of consciousness*," down "the staircase of being, spiral by spiral," away from being, for "being does not lie down below. It is above, always above, or to be more precise, it is in solitary, working thought" (110). So Bachelard ends, not just with nostalgia for the work of science but—here we should remember that he is now seventy-seven years old, unwell, and that he will die a year after the publication of this book—with a remarkable desire to confront once again the constructive difficulties of modern science. The last few paragraphs of *La Flamme d'une chandelle* sum up simply and movingly Bachelard's subversive humanism, his conception of human being as difference, as the work not of life but of thought, and they leave us with a new, final metaphor for human being, that of the worktable. Bachelard's own words draw together the threads of my discussion and are its most fitting conclusion, for through these texts and readings, my wish has been above all to enable him to speak for himself:

> All in all, when the experiences of life are taken into account, experiences that tear one apart and are themselves torn apart, it is in the end when I sit before my blank sheet of paper, before the white page placed on the table at the right distance from my lamp, that I am truly at my *table of existence*.
>
> Yes, it is there at my table of existence that I have known the greatest possible existence, existence as tension—pulled forward and forward again to something beyond, to something above. Peace and tranquillity are all around me; only my being, my being which is in search of being, strains

Gaston Bachelard: Subversive Humanist

forward, driven by the improbable need to be another being, a more-than-being. And so it is with Nothing, with Reveries, that it is believed books can be made.

But when this little album is finished, with its chiaroscuros of a dreamer's mind, a dreamer's psyche, it is once more time to feel nostalgia for severely ordered thought. This candle-romanticism of mine pursued in this book is only half my life at my table of existence. After so much reverie, I am impatient to learn again, to set aside my blank sheet of paper and study a book, study in a book, which is difficult, indeed always rather too difficult for me. And in its tension as it strains forward toward a rigorously developed book, the mind is constructed and reconstructed. The becoming of thought, the future of thought, is in the mind's reconstruction.

Is there, though, still time for me to rediscover the worker I know so very well, and bring him back into this engraving, into my picture of myself? (111–12)

Appendix
Notes
Reference List
Index

Appendix: English Translations of Bachelard's Work

Air and Dreams: An Essay on the Imagination of Movement. Translated by Edith and Frederick Farrell. Dallas: The Dallas Institute of Humanities and Culture Publications, 1988.

Earth and Reveries of Will. Translated by Liliana Zancu. Dallas: The Dallas Institute of Humanities and Culture Publications, forthcoming.

The Flame of a Candle. Translated by Joni Caldwell. Dallas: The Dallas Institute of Humanities and Culture Publications, 1988.

Fragments of a Poetics of Fire. Translated by Kenneth Haltman. Dallas: The Dallas Institute of Humanities and Culture Publications, 1990.

Lautréamont. Translated by Robert Dupree. Dallas: The Dallas Institute of Humanities and Culture Publications, 1984.

The New Scientific Spirit. Translated by Arthur Goldhammer. Boston: Beacon Press, 1985.

On Poetic Imagination and Reverie: Selections from the Works of Gaston Bachelard. Translated with an introduction by Colette Gaudin. Indianapolis: Bobbs-Merrill, 1971.

The Philosophy of No: A Philosophy of the New Scientific Mind. Translated by G. C. Waterston. New York: Orion Press, 1968.

The Poetics of Reverie. Translated by Daniel Russell. New York: Orion Press, 1969. With the title *The Poetics of Reverie: Childhood, Language, and the Cosmos* (Boston: Beacon Press, 1971).

The Poetics of Space. Translated by Maria Jolas. New York: Orion Press, 1964; Boston: Beacon Press, 1969.

The Psychoanalysis of Fire. Translated by Alan C. M. Ross. Boston: Beacon Press, 1964; London: Routledge and Kegan Paul, 1964.

The Right to Dream. Translated by J. A. Underwood. Dallas: The Dallas Institute of Humanities and Culture Publications, 1988.

Water and Dreams: An Essay on the Imagination of Matter. Translated by Edith Farrell. Dallas: The Dallas Institute of Humanities and Culture Publications, 1983.

Notes

CHAPTER 2

1. This description of Heisenberg's uncertainty principle inevitably simplifies, my aim being to convey to nonscientists the general sense of what is involved here. In his book *Physics and Philosophy* Sir James Jeans describes it as follows: "Physics sets before itself the task of coordinating the various sense-data which reach us from the world beyond our sense-organs. If our senses could receive and measure infinitely delicate sense-data, we should be able in principle to form a perfectly precise picture of this outer world. Our senses have limitations of their own, but these can to a large extent be obviated by instrumental aid; telescopes, microscopes, etc. exist to make good the deficiencies of our eyes. But there is a further limitation which no instrumental aid can make good; it arises from the circumstances that we can receive no message from the outer world smaller than that conveyed by the arrival of a complete photon. As these photons are finite chunks of energy, infinite refinement is denied us ... We might think we could avoid this complication by using radiation of infinite wave-length. For the quanta of this radiation have zero energy, and might be expected to provide infinitely sensitive probes with which to explore the outer world. And so they do, so long as we only want to measure energy, but a true picture of the outer world will depend also on the exact measurement of lengths and positions. For this, long-wave quanta are useless ...

". . . in the exact sciences, and above all in physics, subject and object were supposed to be entirely distinct, so that a description of any selected part of the universe could be prepared which would be entirely independent of the observer as well as of the special circumstances surrounding him.

"The theory of relativity (1905) first showed that this cannot be entirely so; the picture which each observer makes of the world is to some degree subjective ...

"The theory of quanta carries us further along the same road. For every observation involves the passage of a complete quantum from the observed object to the observing subject, and a complete quantum constitutes a not negligible coupling between the observer and observed. We can no longer make a sharp division between the two; to try to do so would involve making an arbitrary decision as to the exact point at which the division should be made. Complete objectivity can only be regained by treating observer and observed as parts of a single system . . ." (1942: 141–43).

2. It must be noted that what Bachelard criticizes is the idea of objectivity based on everyday experience. This criticism should not be interpreted as an argument in favor of the subjectivity of modern scientific knowledge. Indeed, he is critical of "subjectivist interpretations" of Heisenberg's principle, stressing its objectivity (*ARPC* 289), together with its realism and its rationalism (296).

CHAPTER 3

1. The periodical *Recherches philosophiques* was founded in Paris in 1931. It set out to encourage new directions in philosophy, and in the wake of Husserl's Paris lectures of 1929 and the publication of *Méditations cartésiennes* (1931), it gave preferential treatment to phenomenology, publishing in its first volume (1931–32) both Heidegger's article and Jean Wahl's preface to *Vers le concret* (1932). Six volumes appeared annually between 1931 and 1936, with the bias toward phenomenology increasingly apparent. Bachelard contributed to volumes 1, 3, 4, and 6, and there are several references in his books to articles published in the periodical, for instance in *La Dialectique de la durée* (1936), where he takes the idea of verticality as the dimension peculiar to human beings from an article by Alexandre Marc in volume 4, adding a footnote reference to an article on duration by Albert Rivaud in volume 3, 94–95. This would seem to confirm the probability of Bachelard's implicit reference to Heidegger's article in *Le Nouvel Esprit scientifique*. There is no evidence of any continuing influence of Heidegger on Bachelard. It is, however, noteworthy that Jean Lescure, commenting on Bachelard's last lectures at the Sorbonne (1953–54) refers to Bachelard's grasp of Heidegger's consciousness of the world as totality as against Husserl's consciousness of a limited object (1983: 200). I have discussed Bachelard's relations with phenomenology in some detail in my doctoral thesis and in articles (see Reference List).

2. Freud, *Five Lectures on Psychoanalysis*, given at Clark University at Worcester, Massachusetts, in September 1909, first published in their English translation in the *American Journal of Psychology* (1910); he describes Breuer's "cathartic treatment" at length in the first lecture, and goes on in the second to outline his own development of the "cathartic procedure." See also Jones (1953–57: Book 1, chap. 11, "The Breuer Period (1882–94)," passim).

CHAPTER 5

1. Dominique Lecourt begins *Bachelard ou le jour et la nuit* (1974) by outlining Bachelard's importance in school and university teaching in France: for many pupils, Bachelard's work is their introduction to epistemology, and his place

on the *agrégation* syllabus confirms and ensures his influence, which, as Lecourt notes, is evident in areas beyond his own, in biology, linguistics, sociology, psychoanalysis, psychology, and law. Lecourt's book is itself a development of lectures he gave on Bachelard in the *agrégation* program (11–14). He refers to *La Formation de l'esprit scientifique* and *La Psychanalyse du feu* as the most widely read of Bachelard's books (121).

2. Michel Serres has discussed this word "formation" in "La Réforme et les sept péchés" (1974: 68–85). His article is a virtuoso performance on a word which is, as he says, "omnidisciplinary" (70). He argues that Bachelard uses it in the moral sense of "reformation." Serres draws attention to an important aspect of the book and of Bachelard's thinking in general, with its notions of pedagogy and catharsis. I have chosen a simpler interpretation here, but one which is, I believe, equally revealing, in order to show how this book differs from its predecessors. It is perhaps salutary for subtle exegetists of Bachelard to note that, in an interview with Alexandre Aspel in 1957, quoted by C. G. Christofides, Bachelard himself referred to the book's title as "ill chosen" (1962: 263–71).

3. This emotionally charged "social dimension" of science is not the same thing as the "social value of objectivation" or the "social aspects of proof" referred to in Chapter 3, Extract I. Whereas Bachelard formerly stressed the "convergence" of minds, he now reflects on their opposition, and on the emotions engendered by the "social life of science."

CHAPTER 6

1. Gagey, Margolin, Smith, and Voisin all argue that Bachelard's work has unity: see for example Margolin (1974: 10, 31, 65, 85); Voisin (1967: 15); Roch Smith concludes that there is "a subtle cross-fertilization" between Bachelard's work on science and on poetry (1982: 135); Gagey begins by describing the "scandal" of a philosopher of science who gave so much time to poetry (1969: 7–10), and goes on to criticize the "hypothesis" of Bachelard's "dimorphism" (139–43), arguing that his work on poetry is both dependent on his epistemology and its completion (269–76). Vadée occupies the middle ground: critical of those who "suspect but do not specify a vague, hidden unity" in Bachelard, he argues "the great coherence of his philosophical thinking" and that contradictions lie not in "this bi-partition" but at the heart of his philosophical theses (1975: 143–44), so that Bachelard's work has coherence because of these contradictions, because of his idealism (152–55). Lecourt had noted these contradictions in Bachelard's epistemology, but reached a different conclusion, namely "the irreducible contradiction between the two aspects of Bachelard's work" (1974: 146–47). Roy shares this view, stressing duality and opposition (1977: 7–8, 11–20, 207–20). Mansuy (1967), Pire (1967), and Therrien (1970) limit their studies of Bachelard to his

work on poetry, which suggests that they regard it as separate from that on science; Pire discusses the question of unity briefly in his conclusion, and adopts what he regards as Bachelard's own view, namely that the two aspects are not to be reconciled (194–98).

2. This article is included in the Gonthier edition (1966) of *L'Intuition de l'instant* (101–11). Lescure describes in *Un Été avec Bachelard* how in 1938, after reading *La Dialectique de la durée*, he invited Bachelard to write this article for the review *Messages* (19–20, 23–24, 41–42); Lescure considers himself to have introduced Bachelard to poetry . . .

3. Lescure's book was the subject of a lawsuit in June and July 1983, brought by Bachelard's daughter and sole heir against the author and the publisher: Lescure had included both unauthorized and private material in the book, which was withdrawn from sale, and reissued with expunctions later in the year.

4. See Christofides (1962), who quotes from Bachelard's interview with Aspel (1957): "I was nearing forty-five when I began teaching courses in literature" (267). See also Therrien (1970: 44, note 1), referring to Bachelard's pleasure in teaching literature to foreign students at Dijon.

5. Barthes goes on to say that "present-day French criticism in its most flourishing aspect can be said to be Bachelardian in inspiration."

6. As regards Bachelard's relations with psychoanalysis, the formative influence is that of Freud, precisely because Bachelard disagrees with him. He will draw on Jung in later books, in particular in and after 1948 (*TRV, TRR, PE, PR*), and although he adapts his ideas, there is greater sympathy and an absence of polemics which in effect diminishes Jung's influence. For this reason, I have chosen not to discuss in this book the question of Bachelard's relations with Jungian theory (these were discussed in my doctoral thesis). Pire (1967: 44–51) and Therrien (1970: 266–70) have touched briefly on this aspect of Bachelard. Bachelard himself said that he had "received Jung too late": see his interview with Aspel (1957), quoted by Christofides (1963: 486).

7. Bachelard may well have been drawn to Lautréamont because of the poet's early interest in mathematics: he came to Paris in order to study this, and there is in his *Chants de Maldoror* a four-page "hymn to mathematics"; Bachelard discusses the question of Lautréamont's "mathematical culture" in chapter 4 of his *Lautréamont*, (89–96).

CHAPTER 7

1. See also Bergson 1961: 96–99 (1910: 129–32).

CHAPTER 8

1. Despite Bachelard's references to the "social value of objectivation," the "social character of proof," and the "convergence" of minds (Chapter 3, Extract I), he does not develop these ideas in *Le Nouvel Esprit scientifique*.

2. This advocacy of "cognito-affective control indispensable for the progress of the scientific mind" was Bachelard's starting point in *La Formation de l'esprit scientifique*, as Chapter 5 has shown (see Extract I). Later in the book (Extract II), he drew attention to the positive role played by "all the emotions involved in the use of reason," interpreting "psychoanalysis" not as a ridding but rather as a revealing of these emotions. The emphasis here is on rivalry rather than cooperation in intellectual relationships. Bachelard's description of the "lone scientist" in this second extract refers only to the subject's relationship with "his own intellectuality" and with objects.

CHAPTER 10

1. Lecourt (1972: 21–26), Margolin (1974: 21), Jean (1983: 105–99), the second half of the book being devoted to Bachelard's "pedagogy of no."

2. Lescure gives a slightly different version in *Un Été avec Bachelard* (1983: 208), adding more of his own reactions to Bachelard's ideas and to his personality.

3. Henri Gouhier made this point at the end of the final discussion at the Cerisy colloquium in 1970, in response to a question about Bachelard's "rejection" of Marxism asked by Richard Medeiros (1974: 192).

Reference List

Bachelard, Gaston. *Essai sur la connaissance approchée*. 1928; Paris: Vrin, 1969. His doctoral thesis, presented in the University of Paris, 1927.

Bachelard, Gaston. *Étude sur l'évolution d'un problème de physique: la propagation thermique dans les solides*. Paris: Vrin, 1928. His complementary thesis, 1927.

Bachelard, Gaston. *La Valeur inductive de la relativité*. Paris: Vrin, 1929.

Bachelard, Gaston. *Le Pluralisme cohérent de la chimie moderne*. Paris: Vrin, 1932.

Bachelard, Gaston. *L'Intuition de l'instant: étude sur la "Siloë" de Gaston Roupnel*. Paris: Stock, 1932. References here are to the current edition (Paris: Gonthier, 1966), which includes Bachelard's article "Instant poétique et instant métaphysique" (1939).

Bachelard, Gaston. *Les Intuitions atomistiques: essai de classification*. Paris: Boivin, 1933. Now published by Vrin.

Bachelard, Gaston. *Le Nouvel Esprit scientifique*. Paris: Alcan, 1934. Edition used: Paris: Presses Universitaires de France, 1973.

Bachelard, Gaston. *La Dialectique de la durée*. Paris: Boivin, 1936. Edition used: Paris: Presses Universitaires de France, 1950.

Bachelard, Gaston. *L'Expérience de l'espace dans la physique contemporaine*. Paris: Presses Universitaires de France, 1937.

Bachelard, Gaston. *La Formation de l'esprit scientifique: contribution à une psychanalyse de la connaissance objective*. 1938; Paris: Vrin, 1972.

Bachelard, Gaston. *La Psychanalyse du feu*. Paris: Gallimard, 1938. Edition used: 1966 paperback in Gallimard's "Collection Idées."

Bachelard, Gaston. *Lautréamont*. 1939; Paris: Corti, 1951.

Bachelard, Gaston. "La Psychanalyse de la connaissance objective." *Annales de l'École des Hautes Études de Gand*, Vol. 3, 3–13. Ghent, 1939.

Bachelard, Gaston. *La Philosophie du non: essai d'une philosophie du nouvel esprit scientifique*. 1940; Paris: Presses Universitaires de France, 1970.

Bachelard, Gaston. *L'Eau et les rêves: essai sur l'imagination de la matière*. Paris: Corti, 1942.

Bachelard, Gaston. *L'Air et les songes: essai sur l'imagination du mouvement*. Paris: Corti, 1943.

Bachelard, Gaston. *La Terre et les rêveries de la volonté: essai sur l'imagination des forces*. Paris: Corti, 1948.

Bachelard, Gaston. *La Terre et les rêveries du repos: essai sur les images de l'intimité.* Paris: Corti, 1948.

Bachelard, Gaston. *Le Rationalisme appliqué.* 1949; Paris: Presses Universitaires de France, 1970.

Bachelard, Gaston. *L'Activité rationaliste de la physique contemporaine.* Paris: Presses Universitaires de France, 1951. Edition used: Paris: Union Générale d'Éditions, 1977.

Bachelard, Gaston. "La Vocation scientifique et l'âme humaine." In *L'Homme devant la science*, an international conference at Geneva, 1952, 11–29. Neuchâtel and Brussels, 1952.

Bachelard, Gaston. *Le Matérialisme rationnel.* 1953; Paris: Presses Universitaires de France, 1972.

Bachelard, Gaston. *La Poétique de l'espace.* 1957; Paris: Presses Universitaires de France, 1964.

Bachelard, Gaston. *La Poétique de la rêverie.* 1960; Paris: Presses Universitaires de France, 1965.

Bachelard, Gaston. *La Flamme d'une chandelle.* 1961; Paris: Presses Universitaires de France, 1964.

Bachelard, Gaston. *Études.* Présentation de Georges Canguilhem. Paris: Vrin, 1970. Collection of articles.

Bachelard, Gaston. *Le Droit de rêver.* Paris: Presses Universitaires de France, 1970. Collection of articles.

Bachelard, Gaston. *L'Engagement rationaliste.* Préface de Georges Canguilhem. Paris: Presses Universitaires de France, 1972. Collection of articles.

Bachelard, Gaston. *Fragments d'une poétique du feu.* Établissement du texte, avant-propos, et notes par Suzanne Bachelard. Paris: Presses Universitaires de France, 1988.

Barreau, Hervé. "Instant et durée chez Bachelard." In *Bachelard: Colloque de Cerisy*, 330–54. Paris: Union Générale d'Éditions, 1974.

Barthes, Roland. "Criticism as Language." *Times Literary Supplement*, 27 September 1963. Rpt. *Essais Critiques* (Paris: Seuil, 1964).

Bergson, Henri. *Essai sur les données immédiates de la conscience.* 1889; Paris: Presses Universitaires de France, 1961. Translated as *Time and Free Will: An Essay on the Immediate Data of Consciousness*, by F. L. Pogson (London: Macmillan, 1910).

Bergson, Henri. *Le Rire.* 1900; Paris: Presses Universitaires de France, 1964. Translated as *Laughter* by Cloudesley Brereton and Fred Rothwell (London: Macmillan, 1913).

Bergson, Henri. *L'Évolution créatrice.* 1907; Paris: Presses Universitaires de France, 1962. Authorized translation *Creative Evolution* by Arthur Mitchell (London: Macmillan, 1919).

Reference List

Bergson, Henri. *L'Energie spirituelle*. 1919; Paris: Presses Universitaires de France, 1976. Translated as *Mind-Energy* by H. Wildon Carr (London: Macmillan, 1920).

Bersani, Leo. "From Bachelard to Barthes." *Partisan Review* 34 (1967): 215–232.

Brée, Germaine, and Zimmerman, E. "Contemporary French Criticism." *Comparative Literature Studies* (College Park) 1, no. 3 (1964).

Canguilhem, Georges. "Sur une épistémologie concordataire." In *Hommage à Bachelard*. Paris: Presses Universitaires de France, 1957.

Canguilhem, Georges. *Idéologie et rationalité dans l'histoire des sciences de la vie: nouvelles études d'histoire et de philosophie des sciences*. Paris: Vrin, 1977.

Caws, Mary Ann. "The *réalisme ouvert* of Bachelard and Breton." *French Review* 37, no. 3 (1964): 302–11.

Caws, Mary Ann. *Surrealism and the Literary Imagination: A Study of Breton and Bachelard*. The Hague: Mouton, 1966.

Champigny, Robert. "Gaston Bachelard." In *Modern French Criticism: From Proust and Valéry to Structuralism*, ed. John K. Simon, 175–91. Chicago: University of Chicago Press, 1972.

Christofides, C. G. "Gaston Bachelard's Phenomenology of the Imagination." *Romanic Review* 52 (1961): 36–47.

Christofides, C. G. "Bachelard's Aesthetic." *Journal of Aesthetics and Art Criticism* 20 (Spring 1962): 263–72.

Christofides, C. G. "Gaston Bachelard and the Imagination of Matter." *Revue Internationale de Philosophie* 17 (1963): 477–91.

Clark, John G. "The Place of Alchemy in Bachelard's Oneiric Criticism." In *The Philosophy and Poetics of Gaston Bachelard*, ed. Mary McAllester, 133–47. Washington, DC: Center for Advanced Research in Phenomenology and University Press of America, 1989.

Cranston, Mechthild. "*Ding* and *Werk*: Heidegger and the Dialectics of Bachelard's *Image*." *Rivista di Letterature Moderne e Comparate* 32 (June 1979a): 130–37.

Cranston, Mechthild. "Voice and Vision, Parole and Regard: Readings of Chagall's Bible by Gaston Bachelard." *French Literature Series* 6 (1979b): 35–47.

Denton, David E. "The Image—Gaston Bachelard's Language of Space." *International Journal of Symbology* 5, no. 5 (1974): 14–21.

Derrida, Jacques. *De la Grammatologie*. Paris: Éditions de Minuit, 1967. Translated as *Of Grammatology* by G. C. Spivak (Baltimore and London: Johns Hopkins University Press, 1974).

Descombes, Vincent. *Modern French Philosophy*. Cambridge: Cambridge University Press, 1980. First published in Paris in 1979 as *Le Même et l'autre: quarante cinq ans de philosophie française (1933–1978)*.

Durand, Gilbert. *Les Structures anthropologiques de l'imaginaire*. Paris: Bordas, 1969.

Egyed, Bela. "Marxism, Science, and Ideology." *Social Praxis* 7 (1980): 91–127.
Ehrmann, Jacques. "Introduction to Gaston Bachelard." *Modern Language Notes* 8 (1966): 572–78.
Elevitch, Bernard. "Gaston Bachelard: The Philosopher as Dreamer." *Dialogue: Canadian Philosophical Review* 7 (1968): 430–48.
Gagey, Jacques. *Gaston Bachelard ou la conversion à l'imaginaire*. Paris: Marcel Rivière, 1969.
Grieder, Alfons. "Gaston Bachelard—'Phénoménologue' of Modern Science." *Journal of the British Society for Phenomenology* 17, no. 2 (May 1986): 107–23.
Grieder, Alfons. "Gaston Bachelard: Phenomenologist of Modern Science." *The Philosophy and Poetics of Gaston Bachelard*, ed. Mary McAllester, 27–53. Washington, DC: Center for Advanced Research in Phenomenology and University Press of America, 1989.
Grimsley, Ronald. "Two Philosophical Views of the Literary Imagination: Sartre and Bachelard." *Comparative Literature Studies* 8 (1971): 42–57.
Hans, James S. "Gaston Bachelard and the Phenomenology of the Reading Consciousness." *Journal of Aesthetics and Art Criticism* 35 (1977): 315–27.
Higonnet, Margaret R. "Bachelard and the Romantic Imagination." *Cahiers Roumains d'Études Littéraires: Revue Trimestrielle de Critique, d'Esthétique, et d'Histoire Littéraires* 1 (1987): 92–109.
Isaacs, Roger M. "Arp and Bachelard: The Fire as Father." *Dada/Surrealism* 2 (1972): 55–66.
Jean, Georges. *Bachelard: L'Enfance et la pédagogie*. Paris: Éditions du Scarabée, 1983.
Jeans, James, Sir. *Physics and Philosophy*. Cambridge: Cambridge University Press, 1942.
Jones, Ernest. *The Life and Work of Sigmund Freud*. London: Hogarth Press, 1953–57.
Kaplan, Edward K. "Bachelard and Buber: From Aesthetics to Religion." *Judaism: A Quarterly Journal of Jewish Life and Thought* 19 (1970): 465–67.
Kaplan, Edward K. "Gaston Bachelard's Philosophy of Imagination: An Introduction." *Philosophy and Phenomenological Research: A Quarterly Journal* 33 (1972): 1–24.
Kaplan, Edward K. "The Writing Cure: Gaston Bachelard on Baudelaire's Ambivalent Harmonies." *Symposium: A Quarterly Journal in Modern Foreign Literatures* 41, no. 4 (Winter 1987–88): 278–91.
Kockelmans, Joseph J. "On the Meaning of Scientific Revolutions." *Philosophical Forum* 11 (1972): 243–64.
Kushner, Eva M. "The Critical Method of Gaston Bachelard." In *Myth and Symbol: Critical Approaches and Applications*, ed. Bernice Slote, 39–50. Lincoln: University of Nebraska Press, 1963.

Reference List

Lafrance, Guy, ed. *Gaston Bachelard*. Ottawa: University of Ottawa Press, 1987.

Lauener, Henri. "Gaston Bachelard and Ferdinand Gonseth: Philosophers of Scientific Dialectics." In *The Philosophy and Poetics of Gaston Bachelard*, ed. Mary McAllester, 55–74. Washington, DC: Center for Advanced Research in Phenomenology and University Press of America, 1989.

Lecourt, Dominique. *L'Épistémologie historique de Gaston Bachelard*. Paris: Vrin, 1969.

Lecourt, Dominique. *Pour une critique de l'épistémologie (Bachelard, Canguilhem, Foucault)*. Paris: Maspero, 1972.

Lecourt, Dominique. *Bachelard ou le jour et la nuit (un essai de matérialisme dialectique)*. Paris: Grasset, 1974.

Lecourt, Dominique. *Marxism and Epistemology: Bachelard, Canguilhem, and Foucault*. Trans. Ben Brewster. London: New Left Books, 1975. This combines 1969 and 1972.

Le Sage, Laurent. "The New French Literary Criticism." *American Society Legion of Honor Magazine* 37, no. 2.

Lescure, Jean. "Paroles de Gaston Bachelard: notes sur les derniers cours de Gaston Bachelard à la Sorbonne." *Mercure de France* 348 (1963).

Lescure, Jean. "Souvenir de Bachelard." In *Bachelard: Colloque de Cerisy*, 226–34. Paris: Union Générale d'Éditions, 1974.

Lescure, Jean. *Un Été avec Bachelard*. Paris: Luneau Ascot, 1983.

McAllester, Mary. "Gaston Bachelard: Towards a Phenomenology of Literature." *Forum for Modern Language Studies* 12, no. 2 (April 1976): 93–104.

McAllester, Mary. "Polemics and Poetics: Bachelard's Conception of the Imagining Consciousness." *Journal of the British Society for Phenomenology* 12 (January 1981a): 3–13.

McAllester, Mary. "The Uses and Abuses of Literary Criticism." *Strathclyde Modern Language Studies* 1 (1981b): 29–42.

McAllester, Mary. "Bachelard Twenty Years On: An Assessment." *Revue de Littérature Comparée* 58, no. 2 (1984): 165–76.

McAllester, Mary. "The Philosophy and Poetics of Gaston Bachelard." In *The Philosophy and Poetics of Gaston Bachelard*, ed. Mary McAllester, 1–12. Washington, DC: Center for Advanced Research in Phenomenology and University Press of America, 1989.

McAllester, Mary. "Unfixing the Subject: Gaston Bachelard and Reading." In *The Philosophy and Poetics of Gaston Bachelard*, ed. Mary McAllester, 149–61. Washington, DC: Center for Advanced Research in Phenomenology and University Press of America, 1989.

McAllester, Mary, ed. *The Philosophy and Poetics of Gaston Bachelard*. Washington, DC: Center for Advanced Research in Phenomenology and University Press of America, 1989.

McAllester, Mary. "On Science, Poetry, and the 'honey of being': Bachelard's Shelley." In *Philosophers' Poets*, ed. David Ward, 153–176. London: Routledge, 1990.

Mansuy, Michel. *Gaston Bachelard et les Éléments*. Paris: Corti, 1967.

Margolin, Jean-Claude. *Bachelard*. Paris: Seuil, 1974.

Margolin, Jean-Claude. "Bachelard and the Refusal of Metaphor." In *The Philosophy and Poetics of Gaston Bachelard*, ed. Mary McAllester, 101–32. Washington, DC: Center for Advanced Research in Phenomenology and University Press of America, 1989.

Mauron, C. *Des Métaphores obsédantes au mythe personnel: introduction à la psychocritique*. Paris: Corti, 1962.

Oxenhandler, Neil. "Ontological Criticism in America and France." *Modern Language Review* 55 (1960): 17–23.

Paris, Jean. "The New French Generation." *American Society Legion of Honor Magazine* 31, no. 1 (1960): 45–55.

Parker, Noël. "Science and Poetry in the Ontology of Human Freedom: Bachelard's Account of the Poetic and the Scientific Imagination." In *The Philosophy and Poetics of Gaston Bachelard*, ed. Mary McAllester, 75–100. Washington, DC: Center for Advanced Research in Phenomenology and University Press of America, 1989.

Peck, Daniel H. "An American Poetics of Space: Applying the Work of Gaston Bachelard." *Missouri Review* 3, no. 3 (Summer 1980): 77–91.

Pire, François. *De l'imagination poétique dans l'oeuvre de Gaston Bachelard*. Paris: Corti, 1967.

Poirier, René. "Autour de Bachelard épistémologue." In *Bachelard: Colloque de Cerisy*, 9–37. Paris: Union Générale d'Éditions, 1974.

Ricardou, J. "Un Étrange Lecteur." In *Les Chemins actuels de la critique: Colloque de Cerisy*, 214–21. Paris: Union Générale d'Éditions, 1968.

Rice, Philip B. "Children of Narcissus: Some Themes of French Speculation (Brunschvicg, Lavelle, Le Senne, Bachelard, Bastide, Canguilhem)." *Kenyon Review* 12, no. 1 (1950): 116–37.

Robinet, André. "Rythme et durée." In *Bachelard: Colloque de Cerisy*, 317–29. Paris: Union Générale d'Éditions, 1974.

Rosny, J.-H. "La Contingence et le déterminisme." *Revue du mois* (Jan. 1914), 22.

Roy, Jean-Pierre. *Bachelard ou le concept contre l'image*. Montreal: Les Presses de l'Université de Montréal, 1977.

Rummens, Jean. "Gaston Bachelard: une bibliographie," *Revue Internationale de Philosophie* 4, no. 56 (1963): 492–504.

Sartre, Jean-Paul. *L'Être et le néant*. Paris: Gallimard, 1943. Translated as *Being and Nothingness*, by Hazel Barnes (London: Methuen, 1957).

Seghers, Pierre. "Letter from Paris." *World Review*, no. 35 (1952): 39–42.

Serres, Michel. "La Réforme et les sept péchés." In *Bachelard: Colloque de Cerisy*, 68–85. Paris: Union Générale d'Éditions, 1974.

Slattery, Dennis Patrick. "Imagining the Stuff of the World: Reflections on Gaston Bachelard and Ivan Illich." *New Orleans Review* 12, no. 3 (Fall 1985): 81–87.

Smith, Colin. "Philosophy in France." *Philosophy* 37, no. 139 (1962): 67–70.

Smith, Colin. "The Role of Reason and the Concept as a Dissimilating Force." *Contemporary French Philosophy: A Study in Norms and Values*. London: Methuen, 1964.

Smith, Colin. "Bachelard in the Context of a Century Of Philosophy of Science." In *The Philosophy and Poetics of Gaston Bachelard*, ed. Mary McAllester, 13–26. Washington, DC: Center for Advanced Research in Phenomenology and University Press of America, 1989.

Smith, Roch C. "Gaston Bachelard and the Power of Poetic Being." *French Literature Series* 4 (1977): 235–38.

Smith, Roch C. "Gaston Bachelard and Critical Discourse: The Philosopher of Science as Reader." *Stanford French Review* 5 (Fall 1981): 217–28.

Smith, Roch C. *Gaston Bachelard*. Boston: Twayne, 1982.

Smith, Roch C. "Bachelard's Logosphere and Derrida's Logocentrism: Is There a Differance?" *French Forum* 10 (May 1985): 225–34.

Snow, C. P. *The Two Cultures and the Scientific Revolution*. Cambridge: Cambridge University Press, 1959.

Stewart, Louis H. "Gaston Bachelard and the Poetics of Reverie." In *The Shaman from Elko: Papers in Honor of Joseph L. Henderson*, ed. Virginia Detloff, Gareth Hall, et al. San Francisco: Jung Institute, 1978.

Sutherland, Fraser. "What We Do with the Dream: Gaston Bachelard and the Materials of the Imagination." *Canadian Fiction Magazine* 39 (1981): 87–96.

Tanner, Tony. "*The Psychology of Fire* by Gaston Bachelard." *London Review of Books* 5, no. 1 (1965).

Therrien, Vincent. *La Révolution de Gaston Bachelard en critique littéraire: ses fondements, ses techniques, sa portée. Du nouvel esprit scientifique à un nouvel esprit littéraire*. Paris: Éditions Klincksieck, 1970.

Tiles, Mary. *Bachelard: Science and Objectivity*. Cambridge and New York: Cambridge University Press, 1985.

Vadée, Michel. *Bachelard ou le nouvel idéalisme épistémologique*. Paris: Éditions sociales, 1975.

Voisin, Marcel. *Bachelard*. Brussels: Éditions Labor, 1967.

Weightman, John. "Day-dreams." *New York Review of Books* 2, no. 6 (1964).

Index

Abraham, Karl, 80
Activité rationaliste de la physique contemporaine, L', 4, 5, 136, 138, 140, 162, 172, 175, 184n2; mentioned, 135
Affectivity: importance of, in scientific mind, 79–80; in teaching science, 85–90
Air: images of, 11, 94; Shelley's imagination of, 122–26. *See also* Elements
Air et les songes, L', 10, 11, 12, 13, 110, 115, 122–26, 168; mentioned, 92
Alchemy: Bachelard's interest in, 79, 80, 90, 132
Althusser, Louis, 4, 5
Applied rationalism, 137–39, 141–42, 143, 151, 153–54; mentioned, 17
Approximate knowledge, 17–21, 21–26, 38; mentioned, 27, 43, 44
Approximation: second-order, 18–20; mathematical, 62
Arendt, Hannah: attitude to science, 171
Aristotle, 48, 60, 62

Bacon, Sir Francis, 48
Balzac, Honoré de, 88
Barreau, Hervé, 31
Barthes, Roland, 93, 186n5
Baudelaire, Charles, 118n1. *See also* Correspondences
Being: and language, 12–13; man's, 13–14, 80, 126, 143, 174, 175–77; metaphors for, 13–14, 176; and time, 32, 34–35; and consciousness, 62, 68, 70, 90; and nothingness, 63–64; openness of, 65; in modern science, 140; philosophers of, 175
Bergson, Henri: Bachelard's subversion of, 5; Bachelard's polemics with, 9–10, 27–28, 31, 37–38, 44, 63–64, 75, 95, 106, 133, 164, 172, 186n1; life force for, 10, 30, 32, 35, 164; and subject, 12; understanding for, 25, 25n6, 61, 95; time and consciousness for, 29–30, 32–33, 32n1, 36–37, 36n2, 38, 63, 158; and nothingness, 63; language for, 133. *See also* Duration; *Homo faber*
Bertaux, Félix, 72, 72n1
Boerhaave, Hermann, 104, 104n6
Bohr, Niels, 50
Bolyai, Janos, 7
Born, Max, 19
Boschère, Jean de, 158n2
Bosco, Henri, 159–60; mentioned, 156, 157, 159n4
Bousquet, Joë: 108–9
Bragg, Sir William Henry, 57, 57nn3,4
Breton, André, 98n2
Brunschvicg, Léon, 21n1, 89, 89n1
Buber, Martin, 39

Canguilhem, Georges, 3, 14
Catharsis, 60, 78–79, 84, 184n2; mentioned, 185n2
Cavaillès, Jean: influence on Bachelard, 135–36
Caws, Mary Ann, 114
Chemistry: Bachelard, teacher of, 29; modern, 29, 45, 55, 58–59, 90, 135, 138; poetry and, compared, 169; "old chemistry books" and imagination of water, 102–3; "naïve chemistry" and imagination of water, 105

197

Child: and world, 17, 25, 27, 99; teaching science to, 29, 79, 80, 83–86, 88, 167–68; rhythmanalysis and, 72; psychoanalysis and, 72, 141
Christofides, C. G., 94, 185n2, 186nn4,6
Claudel, Paul, 71, 107, 107n1
Cogito: Cartesian, Bachelard's critique of, 7, 44–45, 56, 61–62, 64, 67, 139, 148; Bachelardian, 44–46, 65–70, 137–39, 143, 147; compound, 68–69
Condillac, Étienne Bonnot de, 24, 24n5
Consciousness: Bachelard's conception of, 5, 9–10, 12, 16–17, 25, 27–32, 33, 36–38, 61–65, 69, 75, 78, 81, 90, 91, 92, 134, 136–37, 138–39, 141–42, 143–44, 149–51, 158, 169, 172–73, 176; rational, rationalist, 9, 95, 161–62, 165; reader's, 12, 65, 115, 116, 134, 161; child's, 27; of language, 115–16, 134. *See also* Bergson; Husserl
Correspondences, 118, 118n1
Cournot, Antoine-Augustin, 22, 22n2

Dali, Salvador, 104, 104n4
Dante, Alighieri, 125. *See also Divine Comedy*
De Broglie, Louis, 19
Debye, Peter Joseph William, 57, 57n3
Decentering: Bachelard's humanism and, 4, 175; language and, 8; in Bachelard's consciousness of language, 116; poetry and, 163; of subject, 173
Deformation: in Bachelard's epistemology, 96; in material imagination, 103–4, 106
Delcourt, Marie, 39
Depth: in subject, 40, 44–46, 55; as metaphor of human being, 46, 75; vertical time and, 62, 75; imagination of, 107–10
Derrida, Jacques, 4, 5, 13
Descartes, René: Bachelard's subversion of, 5, 7; his polemics with, 44–45, 58–59, 67n1, 153; analysis of wax in *Meditations*, 44–45, 55–56, 55n2; *poêle*, 55n1; *larvatus prodeo*, 148, 148n1. *See also Cogito*
Descombes, Vincent, 3, 4, 80, 166
Description: in science, 17, 21–23, 50; in poetry and art, 23; for idealism, 23

Dialectics: in modern science, 7, 54, 58, 168; in Bachelard's epistemology, 7, 41, 42, 78, 80, 137, 138, 139, 141, 143, 171; in subject, 42, 118, 146, 148–51; non-Cartesian, 58; of duration, 63, 64; of time, 70, 71, 73; in reading, 74; in language, 163; rejection of Hegel's, 171
Dialectique de la durée, La, 9, 61–76; mentioned, 39, 77, 141, 150, 184n1, 186n2
Difference: in Bachelard's epistemology, 6–8, 10, 17, 18, 20, 45–46, 78, 138, 174; Bachelardian *cogito* and, 7; poetry as, 9, 110; reading and, 9, 10, 64–65, 76, 115, 134; as ontological necessity, 10, 45; language and, 13, 134; consciousness as, 30, 62–63; between Bachelard's work on science and on poetry, 91–92, 96, 169; human being as, 176
Dirac, Paul, 19, 169
Discontinuity: epistemological, 16; in mathematics, 29; modern science and, 43, 45, 58–59; in time and consciousness, 62–64, 70, 75, 164, 173; and reading poetry, 64–65, 175; pedagogy of, 168; in ontology, 172; and reading science, 174
Divine Comedy, 169
Dos Santos, Pinheiro. *See* Pinheiro dos Santos
Dream: poetry and, 13, 16, 114; time in Bachelard's dream described, 63, 65–67; problems of language regarding, 95; Lautréamont's biological, 99–100; of water, 102–6, 110, 117–20; of depths, 107; matter and, 120; Shelley's, 124; of tree, 126–28; of smithy, 129–33; reading and, 159–60, 161, 163, 176–77. *See also* Reverie
Droit de rêver, Le, 92, 112, 134
Durand, Gilbert, 93
Duration: Bachelard's conception of, 9, 27, 30, 37–38, 63–64, 67, 70; for Bergson, 29, 30, 31, 32–35; anagenetic, 97, 105–6; as rhythm, 105. *See also* Dialectics

Earth: images of, 11, 92, 94, 102, 105, 108, 117. *See also* Elements

Index

Eau et les rêves, L', 12, 94, 95–96, 102–6, 111, 113, 117–22, 168; mentioned, 92
Einstein, Albert: importance for epistemology of modern science, 4, 6, 21, 50, 51, 77; and Bachelard's theory of time, 28, 31, 33, 34. *See also* Relativity
Elements: law of the four, 94; the four, 94, 96, 97, 102, 105, 153; fifth element postulated, 153
Ellis, Havelock, 80
Eluard, Paul, 98*n*2
Emotion: in poetry and art, 23, 117, 118; rational, 80, 87, 90, 95, 187*n*2. *See also* Affectivity; Catharsis; Psychoanalysis
Empiricism, 141, 145, 150; active empiricism of modern science, 59; and rationalism, 80, 81, 82, 86–87, 151
Engagement rationaliste, L', 5, 6, 7, 8, 29, 61, 76, 77, 135–36, 153, 154, 165, 165–66, 166, 168, 170, 171, 172
Epistemological break, 5, 21, 28, 60, 77, 79, 139
Epistemological obstacle, 77, 79–80, 81–84, 93–94
Epistemology: Bachelard's, 3, 4–7, 15–17, 24, 39–42, 45, 48–54, 60, 77–79, 82–83, 135–36, 145, 145*n*4, 153, 165, 170, 172–73; non-Cartesian, 7, 40, 44, 51, 55, 97; related to conception of consciousness and time, 9, 27, 38, 63, 74; and work on poetry, 91, 93, 96; micro-epistemology, 143. *See also* Approximate knowledge; Knowledge
Error: importance of, in knowledge, 18, 19, 64, 82, 149; in approximate knowledge, 20; disproves idealism, 24; psychological attitudes to, 80–81, 83–84, 85, 145, 151
Essai sur la connaissance approchée, 6, 8, 15–26, 27, 32, 42, 45, 62, 91; mentioned, 28, 136, 172
Études, 7, 28, 29, 169
Existence: and thought, 67–68, 141–42, 146; philosophies of, 75, 137, 139–40, 164; as coexistence, 139–40, 142, 146, 147; superexistence, 146; table of, 176–77. *See also* Heidegger

Expérience de l'espace dans la physique contemporaine, L', 3, 77–78
Experiment: and reason in modern science, 49–52, 54, 137, 138, 140, 147, 151, 166; by modern scientist on Descartes's wax, 57–59

Fabricius, Jean-Albert, 104, 104*n*5
Failure: importance of, 18, 20, 172–73, 174–75; in idea of time, 38; teacher of science lacks sense of, 84, 86
Fire: images of, 11, 101, 102, 117, 129, 133, 156, 159; as epistemological obstacle, 93, 94. *See also* Elements
Flamme d'une chandelle, La, 10, 154–61, 176–77
Form: in material imagination, 97, 102–6; in dynamic imagination, 98–101; isomorphic images and, 107; imagination of, 107; in applied rationalism, 151, 152
Formation de l'esprit scientifique, La, 6, 10, 40, 77–81, 81–90, 92, 93, 137–38, 167, 185*n*1, 187*n*2
Foucault, Michel, 5
Freud, Sigmund, 5, 11–12, 60, 72, 75, 80, 94–95, 138, 141, 184*n*2, 186*n*6. *See also* Superego

Gagey, Jacques, 28, 63, 74, 75, 78, 80, 142, 185*n*1
Geneva: quoted, 136, 165, 167, 170
Gérard-Varet, Louis, 83, 83*n*1
Ghent: quoted, 167, 168
Goblot, Edmond, 49, 49*n*3
Goethe, Johann Wolfgang von, 23, 23*n*3, 131, 131*n*5
Gouhier, Henri, 171, 187*n*3
Guyau, Jean-Marie, 35*n*2

Hegel, Georg Wilhelm Friedrich, 171
Heidegger, Martin, 43–44, 184*n*1
Heisenberg, Werner, 19, 60, 74, 77, 168, 172, 183*n*1, 184*n*2
History of science: and epistemology, 39, 79, 82–83, 87, 88, 152
Homo aleator, 10, 40, 60, 140

Homo faber, 10, 60, 95–96, 105, 106, 140
Homo mathematicus, 140
Hugo, Victor, 103, 103*n*3
Human being: Bachelard's conception of, 10, 60, 75, 94, 97, 110–11, 134, 136, 163, 165, 169, 170, 173, 175; metaphors for, 14, 161, 176. *See also* Man
Human reality, 13, 140, 156, 161, 165
Humanism: Bachelard and, 4–5, 9, 10, 44, 91, 153, 165, 170, 173, 174; subversive humanism, 13, 14, 92, 154, 164–65, 173, 175, 176
Husserl, Edmund: Bachelard and, 5, 9, 12, 137, 139, 140, 153, 161–62, 164–65, 184*n1;* and the concept of intentionality, 9, 139, 161; and *epoche*, 162, 164
Hypothesis: in modern science, 41, 49–50, 140, 148; in reverie, 105, 133; of being, 148

Idealism: humanism and, 4, 9, 44; Bachelard's opposition to, 5, 7, 15, 16–17, 18, 20, 23–24, 27, 31, 45, 75, 137, 170, 172, 185*n1;* discursive, 7, 69; compound, 67
Identity: Bachelard's opposition to, 6, 7, 10, 46, 60, 62, 65, 75–76, 78, 116, 137, 165, 174
Images, poetic: literary, 11, 109; linguistic, 11, 161; written, 11, 161; Bachelard's approach to, 11–12, 64–65, 73, 92, 96–97, 112, 113–14, 175; images and concepts, 91–92; development of interest in, 93–95; compound, 97, 109–10; isomorphic, 97, 107–9; in projective poetry, 101; and language, 109, 115, 156–57; dynamic, 126, 130; material, 127; dynamic and material images of tempering, 129, 133; moral, 131, 133. *See also* Air; Earth; Fire; Water
Imagination: in relation to Bachelard's work on epistemology, 3, 4, 39, 91–92, 96–97, 168–69; open, 11, 112, 119, 121; dynamic, 91, 96–97, 123; material, 91, 95–96, 97, 102–6, 108, 113, 117–18, 127, 130, 132; Bachelard's theory of, 93–94, 97, 109–11, 121, 124, 175; activist, motory, 96–97, 98, 100; *natural, renaturalized,* 99, 120; mesomorphic, 103; formal, 103, 123; viscous, 104; of forces, 106; psychology of, 121
Instant: for Bachelard, 30, 31–32, 33–35, 37, 38, 62, 68, 69, 70; Bergsonian, 158–59
"Instant poétique et instant métaphysique," 92
Intuition de l'instant, L', 14, 27–29, 32–35, 36–38, 74, 90, 95, 115–16, 168; mentioned, 9, 15, 39, 62, 186*n*2
Intuitions atomistiques, Les, 39, 60, 78
Irrational: numbers, 19–20; consciousness of, 38
Isomorphism, 97, 107, 108. *See also* Images

James, William, 18, 46
Jean, Georges, 93, 187*n1*
Jeans, Sir James, 183*n1*
Jones, Ernest, 80, 184*n*2
Joubert, Joseph, 156
Jung, Carl Gustav, 186*n*6

Kalevala. *See* Lönnrot
Kant, Immanuel, 5, 7, 15, 41
Knowledge: human, 3, 61; scientific, 5, 16–17, 21–26, 45, 48, 50, 51, 53, 54, 56, 79, 80, 81–82, 93, 95, 136, 165, 173; effect on knower, 15, 16, 137, 138–39, 143, 146–47, 148–49, 172; for artist and poet, 23; consciousness, time and, 32, 38, 62; (knowledge)[n], 69; teaching scientific, 83–84, 85–89; value of, 167–68; and being, 174. *See also* Approximate knowledge

Lalande, André, 53, 87
Language: man and, 4, 8, 110; in mathematics, 8–9, 50; in poetry, 11, 12–13, 74, 94, 100, 109, 115, 133–34, 156, 161; Bachelard's, 14, 40–44, 75, 95, 96, 116, 157, 166, 168–69; philosopher's inflec-

Index

tion of, 47; reading and, 115, 116, 133–34, 162–63
Laprade, Victor Richard de, 128n3
Laue, Max von, 57, 57n4
Lautréamont, 96–97, 98–110, 112, 114, 115, 116, 163, 164, 168, 169, 186n7; mentioned, 39, 92
Lautréamont, Le comte de (*pseud.* of Isidore Ducasse): Bachelard's reading of his *Les Chants de Maldoror*, 96, 98–101; Lautréamontism, 98
Lautréamont: *Les Chants de Maldoror*, 140–41, 143–49
Lavelle, Louis, 119, 119n2
Lecourt, Dominique, 4, 52, 78, 170–71, 184n1, 187n1
Lescure, Jean, 92–93, 168, 175, 184n1, 186nn2,3, 187n2
Lichtenberger, Henri, 23, 23n3
Life: critique of ordinary, 7, 10, 60, 62, 63, 69, 79, 83, 154, 156–57, 159, 168, 174, 176; antithesis of thought and, 29, 70, 71, 75–76, 91, 96, 97, 125–26, 164, 167, 168, 176; for consciousness and time, 61, 75, 149, 162; poetry as breaking with, 73, 114–16, 141, 156–57, 161, 168, 169
Literary criticism: Bachelard's attitude to, 10–11, 12, 104, 112–14, 131, 134
Lived experience: opposition to, 11, 75, 139, 161–62, 165, 168, 174; poetry and, 156, 175
Lobatchevsky, Nicolai Ivanovich, 7
Lönnrot, Elias: *Kalevala*, 114, 129–30, 129nn1,2, 130n3, 132, 132n7

Mallarmé, Stéphane, 5, 9, 14, 74, 115, 120, 120n3
Man: in modern science and mathematics, 4, 5–10, 47, 89–90, 91, 136, 140, 153–54, 169, 170–71, 173–74; in dynamic and material imagination, 12, 107, 111, 118–19, 124, 125–26, 128, 128n3, 133–34, 160; as unfixed being, 13–14, 163; for Heidegger, 43; and vertical time, 75; unconscious in, 80; and language, 162–63; intellectual destiny of, 167–68. *See also* Homo aleator; Homo faber; Homo mathematicus; Human being
Mansuy, Michel, 96, 185n1
Marc, Alexandre, 75–76, 184n1
Margolin, Jean-Claude, 63, 166, 185n1, 187n1
Marx, Karl, 142, 142n1, 171, 175, 187n3
Marxists: attitude to Bachelard, 4, 18, 166–67, 170–71. *See also* Lecourt; Vadée
Masson-Oursel, Paul, 126, 126n6
Material imagination. *See* Imagination
Materialism: philosophical, 5, 10, 15, 165, 171; dialectical, 19, 42, 170–71; for imagination, 102, 104; modern scientific, 171
Matérialisme rationnel, Le, 5, 8, 135, 137, 138, 140, 153, 171, 173–74
Mathematics: and poetry, 5, 8, 9, 13, 65, 74, 96, 100, 101, 116, 169; in modern science, 6, 7, 8, 19, 28–29, 39, 40–41, 43, 45, 48–50, 55, 59, 60, 89–90, 137, 138, 140–41, 143, 165; as language, 8, 9, 13, 22, 50–51; and metaphysics, 38; teaching of, 85–86, 88
Matter: in modern science, 7, 8, 17, 19, 24, 45, 58–59, 64, 140, 151, 166, 168, 172; imagination and, 12, 94, 95, 97, 102, 105, 106, 107–8, 110, 117, 120, 127, 133–34, 138, 160; for Descartes, 45, 55–56
Mauron, Charles, 114
Metaphor: Bachelard's use of, 13–14, 17, 20, 31, 45–46, 74–75, 87, 107, 161, 174, 176; in poetry, 97, 101, 117, 121, 126, 130; language and, 115, 116
Metaphysics: for Bachelard, 13, 35, 38, 39, 41, 46–48, 50, 52, 55, 58, 59; discursive, 39, 47; compound, 67
Meyerson, Émile, 24, 24n4, 28
Michelet, Jules, 102–3, 103nn1,2
Michelson, Albert Abraham, 34, 51
Microphysics, 7, 18, 19, 21, 40, 45, 55, 74, 77–78

Mind: changed by science, 3, 7, 137, 173, 174, 177; and reality in modern science, 16, 17, 19, 20, 21, 22, 23, 25–26, 28, 39, 46–47, 51, 53, 54, 61; consciousness and, 27; psychology of scientific mind, 40, 41, 48, 52, 53; for Descartes, 44, 55–56; poetry's effect on, 65, 71, 73; and life, 70, 75; formation of, 78–80; prescientific, 79, 104; and epistemological obstacles, 81–82; teaching and, 83–84, 85–88; dialectical powers of, 141, 149–50; relation between two, 143, 145, 146; as value, 167. *See also* New literary mind; New scientific mind

Minkowski, Eugène, 158, 158$n3$
Mirbeau, Octave, 36, 36$n1$
Monakov, Constantin von, 84, 84$n2$, 85
Montaigne, Michel Eyquem de, 173, 174
Mourgue, René, 84, 84$n2$, 85
Music, 30, 31, 36–37

Narcissism, 118, 119, 120
Narcissus, 113, 118–19, 121
New literary mind, 109, 110
Newness: poetry as, 11, 12, 110, 116, 156; modern science as, 28, 50–51, 52, 88, 172; of instants, 37–38; human need for, 110
New scientific mind, 4, 6, 17, 38, 41, 60, 77, 110, 168
Newton, Sir Isaac, 6, 28, 51, 87–88
Nietzsche, Friedrich, 50, 80, 145$n3$, 148$n1$
Nominalism, 48, 86
Non-Euclidean geometry, 6–7, 47$n2$, 50, 51, 169
Nothingness, 36, 38, 63–64
"Noumène et microphysique," 28, 29, 40, 43
Noumenon, 49, 54, 56, 68, 138
Nouvel Esprit philosophique, Le, 6, 7, 8, 9, 10, 39, 40–46, 46–54, 54–59, 60, 74, 75, 78, 111, 168, 169, 184$n1$, 187$n1$; mentioned, 62, 136
Novalis (*pseud.* of Friedrich von Hardenberg), 160

Object: in modern science, 54, 77–78, 86, 92, 97, 140, 154; for reverie and imagination, 95, 103, 120, 126, 155–56, 157–58, 159–60, 161; poetry and, 97, 116; linguistic, 116
Objectivation, 54, 55, 57–59, 89
Objectivity: in modern science, 53–54, 86, 89, 92, 183$n1$, 184$n2$; historian and, 82; social, 85; habits of, 88
Oedipus complex, 71, 72, 88
Oneiric: Bachelard's use of, 95; examples of use, 102, 104, 107, 108, 119, 120, 122–23, 130, 131, 133
Ontology, 10, 32, 39, 45, 64, 134, 148, 172–73
Openness: experience of, 11, 13, 172–73; modern science as, 17, 43; reading poetry and, 65; of language, 134, 162–63
Orpheus complex, 71, 73

Paulhan, Jean, 112
Pauli, Wolfgang, 19, 45, 60, 172
Phenomenology, Husserlian: Bachelard's interest in, 11–12; he modifies it in his epistemology, 137, 140, 150, 161–62, 164; and in his work on imagination, 155, 158–59, 162
Phenomeno-technique, 41, 54, 150
Philosophie du non, La, 97, 136–37, 162, 168, 169, 171, 173; mentioned, 92, 135
Philosophy: Bachelard's place in French, 3–4, 39, 166–67; impact of science on, 4–5, 39, 40–41; language and, 14; criticism of, 21–22, 50, 87, 88–89, 136, 137, 140, 151, 153–54, 164–66, 169, 170, 175; of science, 47–50, 52; his philosophy of science defined, 173
"Philosophy of no," 17, 40, 137
Physics: man and nature in, 5; mathematics and, 22, 29, 42–43, 55; reason in, 45, 49, 51, 52, 150; Descartes's wax for, 57–59; teaching, 83–84; subject and object in, 183$n1$
Pinheiro dos Santos, Lucio Alberto, 64, 71–72

Index

Pluralisme cohérent de la chimie moderne, Le, 15, 39, 169
"Poetic," etymology of, 16
Poétique de la rêverie, La, 91, 92, 95, 155, 162, 163, 169, 176; mentioned, 12, 16, 161, 186n6
Poétique de l'espace, La, 12, 13, 14, 155, 157, 162, 163, 169; mentioned, 161, 186n6
Poetry: for Bachelard, 4, 10, 11, 12–13, 28, 39, 64–65, 71–73, 78, 91–94, 96–97, 110, 120, 125, 135, 138, 152–53, 155, 156–57, 160, 168, 175; as language, 12–13, 100, 115, 156, 161, 162; Lautréamont's, 98–101; primitive, 100–101; projective, 101; and geometry, 101; second-rate, 108, 117; function of, 163. *See also* Mathematics; Reading; Science
Poirier, René, 14, 92
Pope, Alexander, 72, 72n2
Positivism, 49, 151
Possibility: in mathematics and poetry, 9, 13; in science and poetry, 10, 74, 97; in poetry, 11, 156; reality and, 39; in subject, 46, 115, 134, 173, 175; language and, 115, 116, 156, 163; material reverie and, 133; work and, 175
Pragmatism, 18–19, 49, 87, 89, 152, 172, 175
Problem-solving, 138–39, 142–47
Project: Bachelard's conception of, 40, 42–44, 45, 53, 54, 60, 62, 65, 75, 78, 90, 97; projection, of thought, of dreams, 106; natural projection, 106
Projective geometry: discussed, 43; and projective poetry, 96, 101
Psychanalyse du feu, La, 73, 92, 93–94, 95; mentioned, 11, 39, 78, 96, 185n1
Psychoanalysis: criticism of, 11, 12, 75, 94, 99, 112–13, 118–19, 121, 138, 161, 186n6; of objective knowledge, 40, 78, 145, 151; and rhythmanalysis, 64, 72; of reason, 77, 79, 81, 84, 86, 88, 90, 144, 146, 149, 151, 166, 187n2; Bachelard's knowledge of, 80; use of terms, 88–89, 95; material psychoanalysis, 128
Psychologism, 145, 147

Psychology: of scientific mind, 40, 41, 48, 52, 53, 135, 137, 152, 153, 174; epistemology and, 41; of mathematician, 41, 48–49; tautological, 62, 68; of annihilation, 64; compound, 69; epistemological obstacles and, 79, 81, 82–84, 85, 90; of *homo faber*, 105; of mirror, 119; of imagination, 121; exponential, 149, 151; psychological benefit of reading science, 7, 42, 74

Rank, Otto, 80
Rationalism: Bachelard's criticism of, 5, 7, 10, 17, 46, 47–48, 52, 74, 75, 137, 143, 165, 175; his conception of, 52, 150–51, 170; psychology of, 86–87; regional, 138, 165. *See also* Applied rationalism
Rationalisme appliqué, Le, 10, 137–42, 142–47, 147–53, 168; mentioned, 92, 135, 162
Reading: science, 7, 74, 168, 174–75; poetry, 9, 11–13, 64–65, 73–74, 76, 107, 112–16, 131, 134, 156–57, 160, 161, 162, 168, 174–75
Realism: Bachelard's criticism of, 5, 10, 15, 17, 39, 46–48, 49, 59, 74, 151, 175; his modifications of, 41, 42, 47, 48–49, 52, 97, 133
Reality: and mathematics, 6, 8, 28–29, 49, 88–89; and reason in modern science, 7, 9, 17–19, 20–21, 26, 27, 28–29, 41–43, 49, 51–52, 54, 59, 172; approximate knowledge and, 17–20; in modern science, 17, 20, 24, 26, 27, 39, 41, 45, 47–48, 59, 61, 77–78, 81–82, 97, 140, 162, 171; for poetic imagination, 73, 74, 97, 109–10, 129, 156. *See also* Possibility
Realization: scientific reality as, 41, 42, 48, 49–50, 52, 54, 57, 59, 90; imagination as, 99
Reason: revolution in, 5, 6, 7, 8, 75; in modern science, 6–8, 15–16, 20, 54, 59, 82, 136–38, 153, 165–66, 170. *See also* Psychoanalysis; Reality
Recherches philosophiques, 43, 75, 184n1

Rectification: in modern scientific knowledge, 18, 26, 44, 47, 54, 89, 142, 174; in consciousness and time, 62, 141, 149; subjective, 80–81, 89
Relativity, 4, 6, 28–29, 30, 31, 33, 183*n1*
Renéville, Rolland de, 99, 99*n4*, 100, 100*n5*
Repose: Bachelard's concept of, 61–63, 68, 69, 71
Reverie: material, 12, 95, 105–6, 117–18, 122, 128, 130–31, 133–34; precedes action, 25, 27; reading poetry as, 73, 175; defined, 92, 95; tetravalence of, 94; poetic, 100–101, 102, 103, 104, 122; written, 161; mathematical, 169
Rhythm: in time and consciousness, 31, 36, 37, 63, 64, 66, 67, 105; of thought, 56; of mathematical approximation, 62; reading poetry and, 65, 73; for rhythmanalysis, 71, 72
Rhythmanalysis, 64, 71–73, 150
Ribemont-Dessaignes, Georges, 107, 107*n2*
Ricardou, Jean, 114
Riemann, Bernhard, 6, 7
Rilke, Rainer Marie, 71, 123, 123*n4*
Robinet, André, 74
Rodenbach, Georges, 120, 120*n4*
Rosny, J. H., 8
Roupnel, Gaston, 28, 38, 115–16
Roy, Jean-Pierre, 4, 78, 91–92, 114, 185*n1*

Sartre, Jean-Paul, 43, 95, 139, 140, 153, 165
Schopenhauer, Arthur, 41, 69, 125
Schrödinger, Erwin, 19
Science: for Bachelard, 4, 5, 6–7, 15, 18, 21, 22, 24, 57–59, 60, 78, 82, 86, 87, 92, 135, 136, 138, 145, 170, 173; and poetry, 4, 10, 91–92, 92–93, 95–96, 97, 110–11, 141, 152–53, 169–70, 175, 176, 185*n1*; context of, 19; lessons of, 27–28, 30–31, 33; and philosophy, 40–42, 42–44, 44–46, 47, 165; new philosophy of, 46–54; teaching, 83–84, 88; scientific culture, 84, 88, 142, 145, 149, 151, 152, 153, 170; social life of, 89, 185*n3*; man situated in, 136; and humanism, 153–54; specialization in, 171–72
Serres, Michel, 80, 185*n2*
Shelley, Percy Bysshe, 113, 114, 115, 116, 122–26, 122*n1*, 123*nn2,3*
Smith, Colin, 171
Smith, Roch C., 18, 19, 27, 28, 30–31, 38, 74, 185*n1*
Snow, C. P., 169–70
Spinoza, Baruch, 35
Structuralism, 12, 13
Subject: for Bachelard, 4, 16, 20, 27, 41, 42–44, 46, 55–56, 78, 90, 92, 137, 146, 153–54, 155, 163, 173; sovereign, 4, 7, 14, 43, 156–57, 170, 172, 174; rational, 5, 6, 7, 20, 137, 139, 143–44, 146; subject and object as interdependent, 9, 12, 13, 15, 17, 28, 30, 31, 40, 89, 97, 110, 116–17, 134, 143, 150, 155–57, 158, 160–61, 165, 187*n2*; divided thinking, 140–41, 147–49
Substance: metaphysics of, 55–56; for modern science, 58–59; as epistemological obstacle, 79; imagination and, 102–5, 107–8, 132, 160
Superego: Freudian, 141, 151; Bachelardian, 152
Surrationalism, 40

Teaching: Bachelard as teacher, 10–11, 60, 84, 166–68, 170; criticism of science teaching, 29, 83–84, 85–86, 168; epistemological obstacles and, 79; teacher-pupil relations, 80; criticism of literature teaching, 101, 168, 175
Teissier du Cros, Charles, 67–68, 68*n2*
Terre et les rêveries de la volonté, La, 110, 112, 114, 115, 126–28, 129–33; mentioned, 92, 186*n6*
Terre et les rêveries du repos, La, 11, 97, 107–10, 112; mentioned, 92, 186*n6*
Therrien, Vincent, 93, 96, 185*n1*, 186*nn4,6*
Time: and consciousness, 9, 15, 27, 28, 32–35, 36–38, 61–64, 164; vertical, 62–63, 67, 68, 70, 74–75; thought, 64, 71, 73; superimposed, 65–67, 70, 73, 150;

hierarchical, 149–50. *See also* Bergson; Duration; Instant
Transcendence. *See* Heidegger
Trillat, Jean, 58, 58nn5,6
Truth: in science, 18, 50, 53, 59, 88, 94, 145, 165; and error in science, 81, 145; consciousness as, 90; in poetry, 94, 134, 162; socialization of, in science, 142, 145

Unconscious: Bachelard's view of, 11–12, 64, 73, 74, 80, 90, 94–95, 113, 114, 141, 161; geometric, 60; material imagination and, 102, 103, 104, 110

Vadée, Michel, 4, 18–19, 27, 30–31, 78, 167, 170, 171, 185n1
Valéry, Paul, 5, 9, 71, 73, 74, 113, 115, 122, 122n6

Valeur inductive de la relativité, La, 15, 24n4, 28–29
Verification, 18, 21, 53
Verticality, 161, 174, 184n1. *See also* Time

Wagner, Richard, 132, 132n6
Wahl, Jean, 108–9, 109n5, 184n1
War (1939–1945): Bachelard and, 135–36; science and, 153
Water: images of, 11, 94, 102–6, 110, 117–22, 129, 130. *See also* Elements
Woolf, Virginia, 113, 114, 115, 126–28, 133
Wordsworth, William, 72n2
Work: working consciousness, 91; in poetry, 129; concept important in Bachelard's thought, 173–77. *See also Homo faber*